# MOLECULAR APPROACH FOR DETECTION OF WATERBORNE PATHOGENS

# ENVIRONMENTAL SCIENCE, ENGINEERING AND TECHNOLOGY

Additional books in this series can be found on Nova's website under the Series tab.

ENVIRONMENTAL SCIENCE, ENGINEERING AND TECHNOLOGY

# MOLECULAR APPROACH FOR DETECTION OF WATERBORNE PATHOGENS

## ALY E. ABO-AMER

Nova Science Publishers, Inc.
*New York*

### NOTICE TO THE READER
The Publisher has taken reasonable care in the preparation of this book, but makes no expressed or implied warranty of any kind and assumes no responsibility for any errors or omissions. No liability is assumed for incidental or consequential damages in connection with or arising out of information contained in this book. The Publisher shall not be liable for any special, consequential, or exemplary damages resulting, in whole or in part, from the readers' use of, or reliance upon, this material. Any parts of this book based on government reports are so indicated and copyright is claimed for those parts to the extent applicable to compilations of such works.

Independent verification should be sought for any data, advice or recommendations contained in this book. In addition, no responsibility is assumed by the publisher for any injury and/or damage to persons or property arising from any methods, products, instructions, ideas or otherwise contained in this publication.

This publication is designed to provide accurate and authoritative information with regard to the subject matter covered herein. It is sold with the clear understanding that the Publisher is not engaged in rendering legal or any other professional services. If legal or any other expert assistance is required, the services of a competent person should be sought. FROM A DECLARATION OF PARTICIPANTS JOINTLY ADOPTED BY A COMMITTEE OF THE AMERICAN BAR ASSOCIATION AND A COMMITTEE OF PUBLISHERS.

Additional color graphics may be available in the e-book version of this book.

LIBRARY OF CONGRESS CATALOGING-IN-PUBLICATION DATA

Abo-Amer, Aly E.
 Molecular approach for detection of waterborne pathogens / authors, Aly E. Abo-Amer.
     p. cm.
 Includes index.
 ISBN 978-1-61209-572-1 (softcover)
 1. Water--Microbiology. 2. Pathogenic microorganisms--Detection. 3. Drinking water--Health aspects. 4. Drinking water--Microbiology. I. Title.

 QR105.A36 2011
 628.1'6--dc22
                    2011002564

Published by Nova Science Publishers, Inc. † New York

# CONTENTS

# PREFACE

Water is necessary to continue life so a satisfactory quality must be maintained when supplied as drinking water to consumers. Natural water contaminated with faecal matter, domestic and industrial sewage and agricultural waste may result in health risks of disease transmission to humans who use those waters. Many pathogenic microorganisms of human transmitted by water include bacteria and viruses. Microbial pollution of the aquatic environment induces an increased public health risk where it is used as a source of potable water, for fish and shellfish farming, and for recreational activities. Enteric microorganisms from grazing animals can enter a stream in overspill from the grazing lands, and animals with contact to a stream have been shown to drop a part of their daily fecal material directly therein. The risk of infection to humans due to animal fecal pollution is apparent generally to be lower than the risk due to human fecal water pollution. The polymerase chain reaction (PCR) and other biology-based methods, over the past 15 years, have begun to modernize the determination of pathogenic bacteria and viruses in environmental waters. Traditionally, drinking water treatment plants monitor fecal coliforms and other indicator organisms to provide a probable measure of potential fecal contamination and assess efficiency or removal and or inactivation of pathogenic microorganisms. Standard cell culture assay for specific microbial pathogens are not usually performed because these method are expensive, effort intensive, and time-consuming. Molecular biological techniques enable rapid detection of pathogenic microorganisms in water by providing levels of sensitivity and specificity difficult to accomplish with traditional culture-based assays which take days to achieve. Our understanding of the composition, phylogeny, physiology, and functional of microbial communities in the environment have been updated by molecular techniques.

Published applications of molecular methods to drinking water issues include direct detection of pathogens in water, faecal source tracking of either indicator microorganisms or specific pathogens. This book describes the direct detection of pathogens by molecular biologic techniques, i.e. techniques based on the analysis of the nucleic acid content (DNA and RNA) of pathogens. Also, the possibility of genotyping to identify pathogen sources arises, e.g. human versus farm animal.

# Author's Contact Information

**Aly E. Abo-Amer**

Permanent address:
Division of Microbiology, Department of Botany,
Faculty of Science, Sohag University,
Sohag 82524, Egypt.

Present Address:
Division of Microbiology, Department of Biology,
Faculty of Science, University of Taif,
P.O. Box 888, Taif, Saudi Arabia.

*Chapter 1*

# INTRODUCTION

Around the world, the spread of infectious disease in human populations, animal and plant is a serious problem. Water is a common source in the environment of many pathogens affecting these populations. Waterborne pathogens can create a vital problem to drinking water supplies, recreational waters, and water sources for agriculture, and aquaculture, as well as to aquatic ecosystems and biodiversity. Major sources of pathogenic organisms are municipal wastewater effluents, urban runoff, agricultural wastes and wildlife, sewage, sewage derived from hospitals, and insufficiently treated sewage specially from large urban areas representing the greatest threat. Moreover, pathogenic contamination of shellfish beds or irrigation water can cause risks to human food supplies. The World Heath Organization (WHO, 1996) has declared that infectious diseases are the world's sole largest source of human death. Many of these infectious diseases are waterborne and have massive bad impacts in developing countries. Although developed countries have been more successful in controlling waterborne pathogens, water quality problems are still widespread in some countries. Based on U.S.A. estimates, consequently of acute waterborne infections it is possible that about 90,000 cases of illness and 90 deaths happen annually in Canada (ASM, 1999). Thus, there is a necessary to study the consequences of environmental releases of microbial pathogens because there is insufficient knowledge of the sources, occurrence, concentrations, survival and transport of specific microorganisms in the environment. Funding and efforts are required to validate newer molecular biological detection tools, understand the ecology of pathogens in aquatic environments, better predict disease outbreaks, and improve emergency responses. Human bacterial pathogens are considered as an increasing threat to drinking water supplies worldwide because of decreasing quality and quantity of available raw water and the increasing demand of high-quality drinking water. Current advances in molecular biological detection technologies for bacterial pathogens in drinking water carry the promise in improving the safety of drinking water supplies by precise detection and identification of the pathogens. More importantly, the array of molecular approaches allows understanding details of infection routes of waterborne diseases, the effects of changes in drinking water treatment, and management of freshwater resources. Public drinking water supplies were one of the great technological advancements in the developed countries during the 20th century with a significant impact on public health. Today, access to safe drinking water is considered a human right and is currently available for about 83% of the human population with a lack of safe supplies mainly in rural areas of developing countries (WHO, 2004; Abo-Amer et al., 2008). Whereas lack of safe drinking water supply in developing countries is typically because of economic limit, in developed countries health risks by waterborne pathogens may occur through mismanagement of freshwater resources, technological failure and/or improper detection measures (Organization for Economic Cooperation and Development, OECD, 2003). The study of pathogens in drinking water supplies can be seen under different stages such as monitoring of the hygienic quality of the drinking water, assessment of transmission pattern of outbreaks of waterborne infections, identification of sources of microbial contaminations, and understanding the emergence of new microbial pathogens. There is a rather limited knowledge on the microbiological principles controlling the occurrence and pathogenesis of microbial pathogens in drinking water. The most important reason for this lack of knowledge are attributed to that precise detection, identification, and quantification of microorganisms in water are not available and only possible with a combination of classical and molecular methods (OECD, 2003). Currently, bacteria have been identified as the etiological agent in the majority of the waterborne outbreaks in the USA according to the most recent examination data (Liang et al., 2006). Also, in other parts of the developed world the contribution of pathogenic bacteria to waterborne outbreaks is increasing because of changes in life style and the emergence of several new bacterial pathogens (WHO, 2003). Consequently, this chapter concentrated on the analysis of bacterial pathogens in drinking water supplies and referred for the other types of waterborne pathogens to several reviews (Szewzyk et al., 2000; Quintero-Betancourt et al., 2002; Nwachcuku and Gerba, 2004; Fong & Lipp, 2005; Monis et al., 2005; Schets et al., 2008; Giangaspero et al., 2009; Hsu et *al.*, 2009). The greatest challenge is the renewed recognition that resurgent and emerging pathogens with a high resistance to treatment are a significant hazard (OECD, 2003). Additional challenges are more intense land use, climate change, and bioterrorism (Jury and Vaux, 2005; Richardson *et al.*, 2009). Climate change will increase temperature of all open water bulks, which is expected to enhance growth conditions and the survival for some pathogenic bacteria. Additional effects of global warming are increased frequency of heavy rainfalls, storms, and floods, leading to severe problems with runoff, sewage transport, and treatment and cause increased faecal contamination of freshwater resources. Consequently, this chapter focused on demonstration of molecular biological assessments of bacterial pathogens that can meet some of these challenges for safer drinking water supplies.

*Chapter 2*

# WATERBORNE PATHOGEN INDICATORS

Indicator microorganisms have traditionally been used to indicate the presence of pathogens (Berg, 1978). Today, however, we realize many of possible reasons for pathogen presence and indicator absence, or vice versa. Therefore, there is no direct correlation between numbers of any enteric pathogens and indicator (Grabow, 1996). The validity of any indicator system is also affected by the relative rates of removal and destruction of the indicator versus the target risk. So differences due to environmental resistance or even ability to multiply in the environment, all influence their utility. Hence, viral, bacterial, parasitic protozoan and helminth pathogens are unlikely to all behave in the same way as a single indicator group, and certainly not in all situations. Furthermore, viruses and other pathogens are not part of the normal faecal microbiota, but are only excreted by infected individuals. Therefore, the higher the number of people contributing to sewage or faecal contamination, the more likely the presence of a range of pathogens. The occurrence of specific pathogens varies further according to their seasonal occurrence (Berg and Metcalf, 1978; Wilkes, *et al.*, 2009; Richardson· *et al.*, 2009). In this chapter, the WHO and US-EPA lists for waterborne bacterial pathogens of significance for developed countries are considered (Table 1). *Campylobacter jejunic, Campylobacter coli, Helicobacter pylori*, and *Arcobacter butzleri* are considered to be emerging waterborne pathogen according to WHO report (2003). The *Mycobacterium avium* complex and *Helicobacter pylori* are considered because there are several reports of their presence in drinking water habitats, especially in biofilms, and because of their health concerns (WHO, 2003; Watson, 2004). Additionally, *Legionella pneumophila* was added to the list, despite not being strictly waterborne by the oral route, because of its very high health significance in developed countries as demonstrated by many major outbreaks reported over the past years (Marre, 2002; Bartram *et al.*, 2007). Because *Francisella tularensis* (bioweapon agent category A) has been reported several times for being responsible for waterborne outbreaks of tularemia and drinking water supplies represent a potential target for bioterrorism, it is also included. Since there are hundreds of different pathogens that could be present in water, it is generally not economically, technologically, or practically possible to test water to determine whether it contains pathogens. Therefore, the approach has been to use one microorganism (or group of microorganisms) to indicate whether a health risk exists. Different bacterial groups have been used as indicators of the microbiological quality of drinking water, and recreational waters for many years. For example, total coliform bacteria have been used for decades to assess the microbiological quality of drinking water.

Traditionally, hygienic quality and potential health risk from the utilization of drinking water is assessed by the cultivation of indicator bacteria. This classical microbiological method depends on the cultivation of specific bacteria, for example plate counts of coliforms, and has a variety of many drawbacks, like no correlation to many waterborne pathogens and no valid identification of the pathogen (Szewzyk *et al.*, 2000). Bacterial indicator species, like *Escherichia coli* or *Enterococcus faecalis*, go rapidly into a viable but non-culturable (VBNC) state after they are released into freshwater, thereby expressing a completely different set of activities, including virulence traits, than those expressed during normal growth (Lleo, 2005; Brettar and Hofle, 1992). Other bacterial pathogens, like *Legionella pneumophila*, have a complex aquatic life cycle that strongly affects their state of activity (Marre and Bartlett, 2002; Bartram, 2007). Therefore, pathogenesis of bacteria is essentially more complex than that of viruses but also overall activity and cellular metabolism allow more direct analyses of their virulence traits. Owing to these characteristics of waterborne pathogenic bacteria, the current methodology is completely inappropriate for the detection of bacterial pathogens in drinking water and the assessment of their virulence potential. Therefore, advanced detection methodology has to be considered depending on the type of microorganism, the level of taxonomic resolution to be reached, the detection limit to be achieved, and the cost of the analysis per sample. The universal approach currently followed is to analyze nucleic acids, extracted directly out of drinking water, with appropriate of molecular biological methods ranging from PCR and DNA array-based techniques to immunocapturing and fluorescence in situ hybridization (FISH).

## Table 1. Waterborne bacterial pathogens and some molecular tools for their detection

| Pathogenic bacteria | Molecular detection method | References |
|---|---|---|
| *Shigella flexneri* | Real-time PCR | Yang *et al.*, 2007 |
| *Salmonella enterica* | Real-time PCR | Hoorfar *et al.*, 2000 Fey *et al.*, 2004 |
| Pathogenic *Escherichia coli* | Real-time, multiplex, multiplex real-time PCR | Ibekwe *et al.*, 2002; Nguyen *et al.*, 2005; Jothikumar and Griffiths, 2002 |
| *Listeria monocytogenes* | RT-PCR | Klein and Juneja, 1997 |
| *Vibrio cholerae* | Nested, multiplex PCR , multiplex real-time PCR | Singh *et al.*, 2001; Di Pinto *et al.*, 2005; Gubala, 2006, Mendes *et al.*, 2008 |
| *Vibrio* spp. | multiplex real-time PCR | Nordstrom *et al.*, 2007 |
| *Campylobacter jejuni* , *Campylobacter coli* | Real-time, multiplex real-time PCR | Novga *et al.*, 2000 a & b; Hong *et al.*, 2007 |
| *Legionella pneumophila* | Real-time PCR, multiplex real-time PCR | Morio *et al.*, 2007 Templeton *et al.*, 2003 |
| *Mycobacterium avium* complex *Mycobacterium avium* subspecies *paratuberculosis* (MAP) | Real-time PCR | Schonenbrucher *et al.*, 2008 |
| *Francisella tularensis* | Real-time PCR and hybridization probes | Tomaso *et al.*, 2007 |
| *Helicobacter pylori* | Real-time PCR | Nayak and Rose, 2007 |
| *Arcobacter butzleri* | Multiplex PCR | Brightwel *et al.*, 2007 |

*Chapter 3*

# MOLECULAR DETECTION OF BACTERIAL PATHOGENS IN WATER

Traditionally, detection and enumeration of pathogens as well as conventional bacterial indicators have been mainly based on the use of selective culture and standard biochemical methods. However, such methods have a number of drawbacks. First, there is the possibility to miscalculate bacterial density specially when bacteria are physically injured or stressed. For example, slow lactose-fermenting or lactose-negative Enterobacteriaceae, including pathogenic *Salmonella* spp. and *Shigella* spp., could be misestimated by standard coliform test, which leads to false negative conclusions (Ootsubo *et al.*, 2002). Moreover, some microorganisms in the environment are viable, but either fastidious to culture or non-culturable (Kong *et al.*, 2002). Culture-based methods are time-consuming and mono-specific. Because of these reasons, examination of environmental water samples for the presence of pathogens such as *Vibrio cholerae*, *Salmonella* spp., *Shigella* spp., *Campylobacter* spp., *Cryptosporidium* spp. and *Giardia* spp. etc. has not typically become a common procedure. Cultivation-based detection and quantification of bacterial indicators have been used for a long time and are our days' highly standardized procedures for the estimation of the hygienic quality of drinking water (OECD, 2003; Szewzyk *et al.*, 2000). Molecular biological detection and quantification of bacterial pathogens are characterized by a large array of techniques and approaches depending on the pathogen. These approaches include water sampling and concentration of the bacteria, extraction of the nucleic acids, detection and quantification of the pathogen (Bej *et al.*, 1991d; Juck *et al.*, 1996; Iqbal *et al.*, 1997; Tsen *et al.*, 1998) and, in special cases, assessment of virulence of the pathogen. A flow diagram for the molecular analysis of water is given in Figure 1.

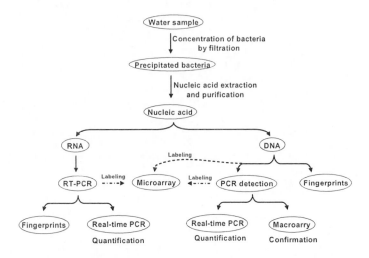

Figure 1. Representation for major steps in the molecular analysis of bacterial pathogens from water. Dotted lines indicate labeling of nucleic acids with fluorochromes after PCR or whole genome amplification.

## A. SAMPLING AND CONCENTRATION OF BACTERIA FROM WATER SAMPLES

Sampling and concentration of bacteria from water for molecular biological analysis are in general achieved by filtration onto membrane filters with a pore size retaining bacteria or concentration of bacteria by ultrafiltration using membranes with appropriate cut-off (Paul and Pichard, 1995). Some freshwater bacteria can pass through membrane filters of a pore size of 0.2 mm depending on the material of the membrane filter (Wang *et al.*, 2007). Therefore, filter sandwiches or depth membrane filters are recommended for complete collection of all bacteria from water samples (Weinbauer *et al.*, 2002). Filtration onto membranes has the advantage of rapid processing of water samples allowing sufficient replicates at low cost. Therefore, membrane filtration is the method of the best choice for routine sampling of small amounts of water (1–5 L as maximum depending on the bacterial densities and the diameter of the filters). Ultrafiltration can concentrate large amounts of water (hundreds of liters) and can include viruses if the proper cut-off is used

(Hill *et al.*, 2007). After the concentration step, samples are frozen as filters or concentrates and can be stored at temperatures below -20 °C until extraction. These frozen samples can be used as a permanent record of the hygienic quality of any water sample if the proper molecular detection assays are accomplished later. This is not possible for cultivation-dependent analyses which can only be carried out with fresh live material.

## B. EXTRACTION OF NUCLEIC ACIDS

Extraction of nucleic acids from frozen samples can be performed with a variety of protocols which are mainly based on a lysis step using enzymes or detergents followed by phenol/chloroform and/or guanidine thiocyanate chemistry for protein elimination and precipitation of the nucleic acids by alcohols (Paul and Pichard, 1995; Sambrook *et al.*, 2001). Commonly, a mechanical step is used to confirm complete lysis of the bacterial cells (Weinbauer *et al.*, 2002). After precipitation, the nucleic acids can be stored below -20 °C for further use. Before starting with the molecular detection, a purification step should be performed in the workflow unless already included in the extraction procedure. This step is necessary to clean environmental nucleic acids from contaminants, such as humic acids and metals, which could inhibit the enzymatic reaction commonly used in the molecular detection technologies. A variety of different methods are used to purify environmental nucleic acids ranging from batch procedures to magnetic beads for high-throughput systems (HTSs) (Sambrook *et al.*, 2001; Kowalchuk *et al.*, 2004; Speers, 2006 ).

## C. DETECTION AND QUANTIFICATION OF THE WATERBORNE PATHOGENS

Purified DNA is used for the molecular detection of one or several targeted pathogens. Generally, this detection is PCR based and involves identification if primers specific for genes of the pathogenic bacteria are used (Table 2). If a single pathogen is detected, for example *Legionella pneumophila*, the assay will provide detection limits and has a particular specificity at a specific taxonomic level. Some primer sets are only specific for the genus *Legionella* which comprises more than seventy species; others are specific for the most relevant pathogenic species that is *Legionella pneumophila* (Bartram, 2007). Primers can target only a single gene, mostly 16S rRNA which has not enough for all pathogenic bacterial species for taxonomic resolution to be species specific, for example for *Shiggella flexeri* (Yang *et al.*, 2007). Therefore, a high-standard molecular assay for drinking water investigation should target two or three genes and include an internal standard to ensure that no inhibition is occurring during the enzymatic reactions needed. To match the results of conventional plate counting, molecular quantification of the pathogen is essential. This is currently performed by real-time PCR with the help of a molecular standard for the targeted pathogen (Fey *et al.*, 2004). Real time PCR analysis of aquatic DNA provides the number of genomes per volume of water for a specific pathogen. This value is not identical with the number of cells that can be estimated by microscopic techniques, such as FISH or immunofluorescence, or the CFUs obtained by plate counting. This is owing to that many bacteria contain more than one genome per cell depending on the growth stage and several bacterial pathogens that can produce mass consisting of several to many cells. Consequently, these factors might lead to different conversion factors from genome copies to cells for different pathogens and need to be determined for each targeted bacterial species individually. Thus, detection of pathogenic bacteria by PCR is often not sufficient to guarantee the specificity needed. A DNA array can be used to confirm the identity of the pathogen and has the advantage of enabling the detection of many target genes in parallel. As demonstrated recently (Miller *et al.*, 2008), medium density chips with a few hundred target genes can cover all relevant waterborne bacterial pathogens. Additional molecular detection tools for pathogenic bacteria are fluorescence in situ hyperdization (FISH) or immunofluorescence microscopy (Kowalchuk *et al.*, 2004). These methods are less applicable to HTS detection but are very useful for understanding mechanisms of survival and infection on a cellular level (Braganca *et al.*, 2007). All these advanced molecular biological detection methods require a high degree of expertise and sophisticated technology and are far from being applied for the routine use for drinking water analysis. Therefore, there is a massive need for technological improvement, standardization, validation, and automation of the molecular detection technology. Validation of molecular detection methods for the environmental detection of microorganisms is a very recent field and highly recommended for quality control of drinking water (OECD, 2003).

## Table 2. Some evaluated primers for detection of waterborne pathogens

| Organism | Target gene | Forward (F) and reverse (R) primers (sequence 5´→3´) | References |
|---|---|---|---|
| *Escherichia coli* | *lamB* | F: CTGATCGAATGGCTGCCAGGCTCC<br>R: CAACCAGACGATAGTTATCACGCA | Kapley *et al.*, 2000a, Bej *et al.*, 1990 |
| | *dcuS* | F: TAATTATCGATGACGACGCAATGG<br>R: CTAGCTCGTCGGTTGA AAAT | Abo-Amer *et al.*, 2008 |
| | *dcuR* | F: ACAAATGGGTAGATCAGATTAATCT<br>R: ACATGTGTGAACCCTCGCGA | Abo-Amer *et al.*, 2008 |
| | *groEL* | F: CCGTGGCTACCTGTCTCCTTACTT<br>R: CCAGCAACCACGCCTTCTTCTACC | Sheridan *et al.*, 1998 |
| | *rpoH* | F: CCACAGGCGGATTTGATTC<br>R: GGTTTGCCGCCTGCTCTTC | Sheridan *et al.*, 1998 |
| | *tufA* | F: ACTTCCCGGGCGACGACACTC<br>R: CGCCCGGCATTACCATCTCTAC | Sheridan *et al.*, 1998 |
| | *uidA* | F: TTGCTGTGCCAGGCAGTTT<br>R: ATCATGGAAGTAAGACTGC | Lleo *et al.*, 2005 |
| *E.coli*: ETEC, EHEC | *Slt-1* | F: TTTACGATAGACTTCTCGAC<br>R: CACATATAAATTATTTCGCTC | Lleo *et al.*, 2005 |
| *E.coli*: EHEC, EPEC | *eaeA* | F: TATTATGCTTAGTGCTGG<br>R: CGGAATCATAGAACGGTA | Lleo *et al.*, 2005 |
| *Enterococcus faecalis* | *pbp5* | F: CATGCGCAATTAATCGG<br>R: CATAGCCTGTCGCAAAAC | Lleo *et al.* 1999 |
| *Salmonella* | *invA* | F: CCTGATCGCACTGAATATCGTACTG<br>R: GACCATCACCAATGGTCAGCAGG | Kapley *et al.*, 2000b |
| | *phoE* | F: AGCGCCGCGGTACGGGCGATAAA<br>R: ATCATCGTCATTAATGCCTAACGT | Kapley *et al.*, 2001 |
| | *spvA* | F: TGTATGTTGATACTAAATCC<br>R: CTGTCATGCAGTAACCAG | Kapley *et al.*, 2001 |
| | *spvB* | F: ATGAATATGAATCAGACCACC<br>R: GGCGTATAGTCGGCGGTTTTC | Kapley *et al.*, 2001 |
| | *IpaB* | F: GGACTTTTTAAAAGCGGCGG<br>R: GCCTCTCCCAGAGCCGTCTGG | Kong *et al.*, 2002 |
| *Shigella* spp. | *IpaH* | F: CCTTGACCGCCTTTCCGATA<br>R: CAGCCACCCTCTGAGGTACT | Kong *et al.*, 2002 |
| *Aeromonas* spp. | *Aero* | F: TGTCGGSGATGACATGGAYGTG<br>R: CCAGTTCCAGTCCCACCACTTCA | Kong *et al.*, 2002 |
| *Yersinia enterocolitica* | *Ail* | F: CTATTGGTTATGCGCAAAGC<br>R: TGGAAGTGGGTTGAATTGC | Kong *et al.*, 2002 |
| *Vibrio* | *ctxA* | F: CTCAGACGGGATTTGTTAGGCACG<br>R: GATCTTGGAGCATTCCCACAACC | Kapley *et al.*, 2000b |
| *Vibrio parahaemolyticus* | *Vpara* | F: GCTGACAAAACAACAATTTATTGTT<br>R: GGAGTTTCGAGTTGATGAAC | Kong *et al.*, 2002 |
| *Vibrio cholerae* | *rtxA* | F: AGCAAGAGCATTGTTGTTCCTACC<br>R: ACTTCCCTGTACCGCACTTAGAC | Gubala, 2006 |
| | *epsM* | F: TGGTTGATCGCTTGGCGCATC<br>R: ATGGCAGCCTTTGAGTGAG | Gubala, 2006 |
| | *mshA* | F: ACACCTGGAACAGTTATTGATGGC<br>R: TCACTCGAAGTATCTAGCGTTTGC | Gubala, 2006 |
| | *tcpA* | F: TGCAATGACACAAACTTATCGTAG<br>R: CCCATAGCTGTACCAGTGAAAG | Gubala, 2006 |
| | *EpsM* | F: GAATTATTGGCTCCTGTGCAGG<br>R: ATCGCTTGGCGCATCACTGCCC | Kong *et al.*, 2002 |

Molecular techniques can also be used for the detection of faecal contaminations in general and not only for the detection of a specific pathogen. The microbial source tracking (MST) is also highly relevant for the assessment of human health risks and for the identification of the route of infection and its prevention (Simpson *et al.*, 2002). Currently, several approaches are followed such as speciation-finding microbial species indicative of the source, biochemical marker substances such as faecal sterols, comparison of two or more faecal indicator ratios, and DNA fingerprints using environmental DNA and specific phylogenetic genes of faecal bacteria (Simpson *et al.*, 2002). MST is a rapidly developing field in environmental and food microbiology, and will allow differential between faecal contaminations from human or animal sources and reveal details of infection routes for waterborne diseases (Blanch *et al.*, 2006).

## 3.1. POLYMERASE CHAIN REACTION (PCR)

The polymerase chain reaction is a technique which produces multiple copies of the DNA of interest. The details of PCR procedure can be obtained from a variety of sources (Bej *et al.*, 1990; Waage *et al.*, 1999a & b; Burtscher *et al.*, 1999, Sambrook *et al.*, 2001). Briefly, the PCR method uses a thermostable DNA polymerase enzyme (*Taq* DNA polymerase) to create multiple copies of target DNA. Detection of target DNA, for example in the genome of a viral or bacterial pathogen, is achieved through the use of synthetic single stranded DNA called as oligonucleotide primers (Table 2). These primers can be designed to be specific for an individual organism, or for a group of organisms. PCR is run by using a cycling of three different temperatures. Double stranded DNA is separated into individual strands using a high temperature, commonly over 90 $^{\circ}$C. A lower temperature is then used to anneal the primers to the target section of DNA. At an intermediate temperature between the previous two temperatures, the DNA polymerase produces a mirror copy of the target DNA. This cycle of temperatures is then repeated multiple times. Over 25-30 cycles, due to the exponential nature of the PCR method, more than $10^9$ copies of the target DNA can be theoretically produced (Figure 2). This large number of a target DNA segment can then be detected using standard detection methods such as agarose gel electrophoresis or membrane hybridization. The ability of PCR to produce extremely large numbers of copies of a specific nucleic acid segment provides the requirements for the rapid, very sensitive and specific detection of desired microorganisms in a water sample. In addition, PCR can be employed as a standard method or modified to semi-nested PCR or nested PCR methods (Gajardo *et al.*, 1995; Le Guyader *et al.*, 1995; Straub *et al.*, 1995; Mayer and Palmer, 1996). Semi-nested and nested PCR, where a second PCR reaction is performed using additional primers, increases detection efficiencies through the further amplification of amplified DNA. These modified methods work through the use of one or both of the original primers (semi-nested) or through a completely different set of more selective primers (nested). Both of these modified methods have been demonstrated to significantly improve the detection efficiency of the PCR method (Gajardo *et al.*, 1995; Straub *et al.*, 1995). Sensitivity of PCR methods have also been increased through the use of membrane hybridization detection of PCR products with specific DNA probes (Hay *et al.*, 1995; Schwab *et al.*, 1995; Laberge *et al.*, 1996) or by using enzyme-linked immunoassay (ELISA) (Ritzler and Altwegg, 1996). Because it is highly possible that water and wastewater samples contain more than one microbial pathogen, multiplex PCR can be used to detect more than one target in a single PCR reaction (Way *et al.*, 1993; Pepper *et al.*, 1997; Picone *et al.*, 1997; Rochelle *et al.*, 1997). Multiplex PCR includes the use of a number of DNA primers, each of which are designed to detect specific microbial species in a single PCR reaction (Picone *et al.*, 1997).

### 3.1.1. Detection of Waterborne Pathogens by PCR

Because of increasing population and industrialization, providing safe drinking water is one of the challenges faced by most countries. Unsafe drinking water and insufficient hygiene are one of the reasons for more than three million people die annually from water-related diseases (WHO, 2001). The list of enteric bacteria is increasing with emerging pathogens that have been recognized to be associated with drinking water. Many of these pathogens are not actually new but were not previously identified owing to lack of detection methods. For example, *Campylobacter jejuni* was considered to be a rare opportunistic pathogen causing blood stream infection until it was recognized as a cause of diarrhea, and a possible infectious agent in drinking water (Szewzyket *et al.*, 2000). In 1892, *Escherichia coli* was suggested as an indicator bacterium for water monitoring. Over time the approaches to analysis have changed but the coliforms still remain markers in drinking water safety measures. Although this has its advantage, there is an increasing need to recognize new markers, as discussed that the coliforms alone cannot function as reliable markers for enteric pathogens (Leclere et al., 2001). This is mainly owing to analysis design and the cross reactivity shown by biochemical tests. Another limitation to this type of monitoring is the time required for arriving at conclusions. These shortages can be overcome by using molecular tools, rather than conventional tools, for monitoring water. The molecular approach will ensure the efficiency of detection and reduce the time involved, consequently, helping in controlling the problem of unsafe drinking water. The use of molecular tools in diagnostic applications has been reviewed (Abramowitz, 1996).

Application of molecular-based methods has motivated the interest in directly monitoring for specific pathogens in surface waters (Scott et al., 2002; Simpson et al., 2002). The mainly tool of most molecular-based techniques is the use of polymerase chain reaction (PCR) to evaluate the presence of selected pathogens by detection of its specific pathogenic genes or small-subunit ribosomal RNA (SSU rRNA). PCR-based technology could detect unculturable microorganisms or microorganisms that do not easily grow under laboratory conditions. It is important to notice that the molecular-based techniques could be also used to detect conventional faecal indicators including *Escherichia coli* (Bej et al., 1990, 1991a). Therefore, PCR has become a practical molecular based detection alternative to culturing, microscopy and biochemical testing for the identification of bacterial species. Application of the PCR-based methods also detected the presence of *Salmonella* spp. (Fukushima et al., 2002; Malorny et al., 2003; Lofstrom et al., 2004), *Shigella* spp. (Fukushima et al., 2002; Horman et al., 2004), *Campylobacter* spp. (Lubeck et al., 2003a & b), *Legionellae* (Wellinghausen et al., 2001), *Vibrio vulnificus* (Panicker et al., 2004a), different pathogenic strains of *Escherichia coli* (Franck et al., 1998), protozoan parasites and enteroviruses (Fout et al., 2003; Haramoto et al., 2004; Horman et al., 2004) in various environmental waters. Moreover, there have been several reports of the application of PCR for detecting *V. cholera* focusing on designing assays detecting single gene targets, providing information as to the presence of virulence and regulatory genes, sequences encoding outer membrane proteins and genes involved in O-antigen biosynthesis (Shirai et al., 1991, Fields et al., 1992, Chowdhury et al., 1994; Albert et al., 1997; Nandi et al., 2000; Chow et al., 2001; Lipp et al., 2003). It is estimated that there are >40 000 references to PCR that describe its employment in various applications (White, 1996).

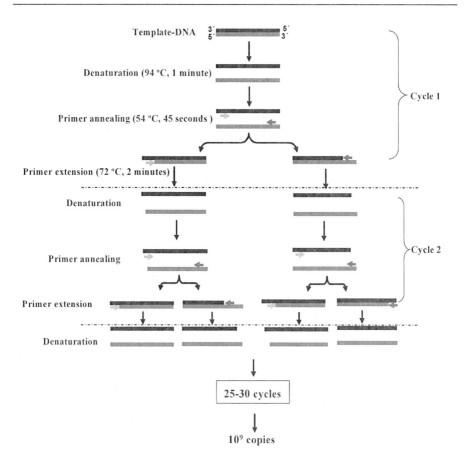

Figure 2. Schematic representation of principle of the polymerase chain reaction (PCR).

However, PCR technology continues to develop at such a fast rate that other new techniques are sure to be discovered. The bottleneck in getting the PCR as diagnostic tool from the laboratory to the field could be focused if a case-specific solution is found (Abramowitz, 1996). The first step is to carefully plan the logistics of handling the samples. One of the interests about the use of PCR for water monitoring is the recovery of PCR-compatible DNA. Although bacterial harvesting methods need to be improved because they still involve filtration and centrifugation, a novel method for template preparation has been reported. This involves only a lysis step, for collected residues on filter paper, with 0.5 N NaOH and neutralization with 1 M Tris, pH 7.5 to yield a PCR compatible template (Kapley et al., 2000a). Similarly, inclusion of fluorescent dyes to the DNA during amplification has led to new advances in detection methods. This has led to designing a new 'kinetic' thermocycler in which the products of a PCR reaction can be monitored at each cycle. Moreover, an Advanced Nucleic Acid Analyzer has been reported that can detect bacteria within 7 minutes (Belgrader et al., 1999). This battery-operated device can be utilized in the field and has software that can be used by first timers. New issues are emerging in measures in drinking water quality as a result of the availability of molecular tools (Szewzyk et al., 2000; Leclere et al., 2001). For example, Escherichia coli could still be considered as an indicator bacterium for feacal contamination if the sample is concentrated to DNA and analyzed by the lamB-specific locus (Bej et al., 1990 Bej et al., 1990; Kapley et al., 2000a). At the same time the DNA sample could be further analyzed for wide range of target loci including the emerging new pathogens. With the initiation of real time PCR, a quantitative option for amplified DNA is possible. This could be further supported by multiplex PCR with a choice of fluorescent dyes, which promises future in quantitative estimation of target DNA copies.

However, interpreting the results from a PCR assessment is more complicated than simple conclusion about presence or absence of pathogens. A positive result definitely provides an indication of pathogen contamination, while a negative result is very difficult to interpret without knowing the precise detection limit for the assay (Loge et al., 2002). Detection limits depend on series of factors including volume of filtered water sample (Loge et al., 2002), the efficiency of nucleic acid extraction (Head et al., 1998), the presence of inhibitory compounds in the PCR reaction (Loge et al., 2002) and formation of chimeric PCR products (Kreader, 1996). The PCR efficiency also depends on chosen PCR conditions and specificity of primer sets. To increase the detection probability of pathogens, pre-culturing was sometimes conducted. This step, however, does not allow quantitative measurements of pathogens; only presence or absence of pathogens can be evaluated. Significantly, DNA-based PCR methods do not show the ability to differentiate between viable and non-viable organisms since DNA of both live and dead cells can be amplified.

In this concern, detection of SSU rRNA seems to be a promising tool for both detection of certain microorganisms and estimation of their viability in the environments since the copy numbers of it have certain relationship to the metabolic activity. For this reason, reverse transcription PCR (Burtscher and Wuertz, 2003), real time RT-PCR (Fey et al., 2004; Gibson et al., 1996), multiplex RT-PCR, fluorescence in situ hybridazation (FISH) (Amann et al.,

1995) and DNA microarray techniques (Chizhikov *et al.*, 2001) have received increasing attentions. Application of these methods seems to be promising as a tool for detection and quantification of viable microorganisms in the studies where survival ability of specific organisms is under special interest. For example, quantitative RT-PCR has been successfully used for quantification of the metabolically active *Salmonella* spp. in different environmental water samples (Fey *et al.*, 2004).

## 3.2. MULTIPLEX POLYMERASE CHAIN REACTION (M-PCR)

PCR is one such established molecular technique and the efficiency of detection can be increased by designing a duplex PCR, in which, two pathogens or loci can be detected in a single PCR reaction (Kapley *et al.*, 2000a; Kapley and Purohit, 2001). The introduction of a multi-step thermocycling program can increase the specificity as demonstrated for the simultaneous detection of three target loci in a single reaction (Kapley *et al.*, 2000a). *E. coli, Salmonella* and *Vibrio* are the most reported water-borne pathogens (Szewzyk *et al.*, 2000), and the primer sets used to monitor these pathogens by PCR are listed in Table 2. *E. coli,* an indicator bacterium, can be monitored by amplifying a PCR product using the *lamB* gene (Bej *et al.*, 1990) and *Salmonella* could be monitored using the *invA* locus (Kapley *et al.*, 2000), which has also been demonstrated in river water samples (Kapley *et al.*, 2001). Both the *lamB* and *invA* loci encode for the bacteriophage specific surface proteins. For monitoring *Vibrio*, a toxin encoding locus *ctxA*, has been used (Kapley *et al.*, 2000b; Kapley and Purohit, 2001). The primers for these loci were also evaluated for spiked drinking water and raw water samples (Bej *et al.*, 1990; Kapley and Purohit, 2001; Kapley *et al.*, 2001; Kapley *et al.*, 2000b). Recently, the use of nucleic acid probes and PCR have provided highly sensitive detection methods for specific pathogens in environmental samples (Bej *et al.*, 1991b).

Multiplex PCR system has been rather established as a rapid and highly sensitive technique for simultaneous detection of many microorganisms in a single PCR tube (Franck *et al.*, 1998; Fukushima *et al.*, 2002). Clinical isolates of *Vibrio cholera* and those from the environments can be differentiated on the basis of multiplex PCR for the presence of *Vibrio cholera* O1, O139, non-O1, and non-O139 strains (Singh *et al.*, 2001).

The use of the multiplex PCR system has provided rapid and highly sensitive methods for the specific detection of pathogenic microorganisms in the aquatic environment. Most multiplex PCR assays for pathogen detection have focused on only one (Shangkuan *et al.*, 1995; Kong *et al.*, 1999), two (Way *et al.*, 1993) or three (Kong *et al.*, 1995) different types of organisms.

Multiplex PCR assay was developed for the detection of three organisms in ballast water (Aridgides *et al.*, 2004), the reverse transcription assay for the simultaneous detection of a gene from each of *Vibrio cholerae*, *Escherichia coli* O157:H7 and *Salmonella typhi* (Morin *et al.*, 2004). In a similar manner, a recently published assay describes the differentiation of three closely related *Vibrio* spp. using a threeplex PCR (Di Pinto *et al.*, 2005). This work described specific species differentiation of *Vibrio alginolyticus*, *Vibrio cholerae* and *Vibrio parahaemolyticus* by targeting the differences in the one collagenase gene between each of the species.

Recent study reported the development of a multiplex PCR method that permits the simultaneous detection of six different types of waterborne pathogens like *Aeromonas* spp., *Salmonella* spp., *Shigella* spp., *Vibrio cholerae*, *Vibrio parahaemolyticus* and *Yersinia enterocolitica* in a single PCR tube (Kong *et al.*, 2002). The target genes (Table 2) used in this study were the aerolysin (*aero*) gene of *Aeromonas hydrophila*, the invasion plasmid antigen H (*ipaH*) gene of *Shigella flexneri*, the attachment invasion locus (*ail*) gene of *Yersinia enterocolitca*, the invasion plasmid antigen B (*ipaB*) gene of *Salmonella typhimurium*, the enterotoxin extracellular secretion protein (*epsM*) gene of *Vibrio cholerae* and a species-specific region of the 16S–23S rDNA (*Vpara*) gene of *Vibrio parahaemolyticus*. Multiplex PCR using the six pairs of primers produced specific amplicons of the expected sizes from mixed populations of reference bacterial strains in seawater and from pure cultures. The multiplex PCR assay was specific and rapid, with a turnaround time of <12 h. The detection limit of the assay for the bacterial targets was estimated at $10^0$–$10^2$ cfu. Both laboratory and field validation results demonstrated that the multiplex PCR assay developed could provide a cost-effective and informative supplement to traditional microbiological methods for routine monitoring and risk assessment of water quality (Kong *et al.*, 2002).

Particularly, the developed techniques of the detection of more than one gene from the same microorganism increase specificity and accuracy of identification. Several multiplex assays specific for the detection of only *Vibrio cholerae*, targeting 2 or more genes, have been reported (Keasler and Hall, 1993; Shangkuan *et al.*, 1995; Hoshino *et al.*, 1998; Kapley and Purohit, 2001; Rivera *et al.*, 2001; Singh *et al.*, 2001; Rivera *et al.*, 2003). Furthermore, multiplex PCR assays have also been successfully applied for the characterization and differentiation of different *Vibrio cholerae* strains and isolates (Studer and Candrian, 2000, De *et al.*, 2001; Mitra *et al.*, 2001; Singh *et al.*, 2001). A PCR assay was also used to detect *Vibrio cholerae* by amplified products which was detected through their hybridisation to specific probes bound to a DNA microarray (Panicker *et al.*, 2004a). The multiplex assay was utilized to detect four previously described targets (Rivera *et al.*, 2001) as well as two targets of *Vibrio vulnificus*, and three targets of *Vibrio. parahaemolyticus*, resulting in a high specificity of detection.

The ability to rapidly monitor for various types of microbial pathogens would be extremely useful not only for routine assessment of water quality to protect public health, but also allow effective assessments of water treatment processes to be made by permitting pre- and post-treatment waters to be rapidly analyzed. However, several post-PCR steps for detection of the organisms, including nick translation, biotinylation and hybridization of the PCR products to the DNA arrays, are time consuming. Thus, the high throughput and cost-effective multiplex PCR system developed could provide a powerful supplement to conventional methods for more accurate risk assessment and monitoring of pathogenic bacteria in the environment.

## 3.3. REAL-TIME POLYMERASE CHAIN REACTION (Q-PCR)

However, since the majority of the published assays employs traditional PCR methods which means that they require product characterization by gel electrophoresis and are time consuming and hard task. Therefore, Real-time PCR has the potential to provide a quicker and more sensitive method for the detection of a wide range of microorganisms. Quantitative real-time PCR using the Perkin Elmer/Applied Biosystems (PE/ABD) (Foster City, CA, USA) Prism 7700 Sequence Detector System is a relatively new technology that provides a broad dynamic range (at least five orders of magnitude) for detecting specific gene sequences with excellent sensitivity and precision (Gibson *et al.*, 1996; Heid *et al.*, 1996). DNA and RNA can be quantified using this detection system without difficult post-PCR processing. This technology can be performed in a core facility environment because of the costs associated with instrument purchase, setup, and maintenance. Quantitative real-time PCR using the PE/ABD 7700 is based on detection of a fluorescent signal produced proportionally during the amplification of a PCR product. The chemistry is the key to the detection system (Figure 3). A probe such as TaqMan and molecular beacons are designed to anneal to the target sequence between the traditional forward and reverse primers. TaqMan probes and molecular beacons are oligonucleotides that are conjugated to reporter and quencher dyes at 5′ and 3′ ends, respectively. In an intact probe or a beacon, the quencher dye suppresses the fluorescence emission of the reporter dye. However, the modification of a TaqMan probe (hydrolysis by Taq polymerase to cleave reporter moiety from the probe) or a conformational change in a molecular beacon during annealing and extension phases of the PCR process results in an increase in the reporter dye's fluorescence intensity. The continuous measurement of incremental fluorescence increase at each PCR cycle provides an accurate estimate of the number of cells of a bacterial pathogen present in a contaminated food, contaminated water or feacal sample. The probe is labeled at the 5' end with a reporter fluorochrome (usually 6-carboxyfluorescein [6-FAM]) and a quencher fluorochrome (6-carboxy-tetramethyl-rhodamine [TAMRA]) added at any T position or at the 3' end (Livak $et$ $al$, 1995). The probe is designed to have a higher $T_m$ than the primers, and during the extension phase, the probe must be 100% hybridized for success of the assay. As long as both fluorochromes are on the probe, the quencher molecule stops all fluorescence by the reporter. However, as $Taq$ DNA polymerase extends the primer, the intrinsic 5' to 3' nuclease activity of $Taq$ degrades the probe, releasing the reporter fluorochrome (Holland $et$ $al.$, 1991). The amount of fluorescence released during the amplification cycle is proportional to the amount of product generated in each cycle.

Moreover, in real-time PCR amplification, the products can be detected through the use of a fluorescent double stranded DNA binding dye such as SYBR Green I, and the products are differentiated by analysis of their melting temperature. This has been reported for the detection of various bacterial species including $Escherichia$ $coli$ O157:H1, $Salmonella$, $Listeria$ and $Vibrio$ $vulnificus$ (Jothikumar and Griffiths, 2002; Jothikumar $et$ $al.$, 2003; Panicker $et$ $al.$, 2004; Wang $et$ $al.$, 2004). Real-time PCR technique, which uses fluorescence-based detection of target gene sequence, allows real time quantitative monitoring of microorganisms of interest (Heid $et$ $al.$, 1996; Lyons $et$ $al.$, 2000; Novga $et$ $al.$, 2000b; Bellin $et$ $al.$, 2001; Hein $et$ $al.$, 2001; Huijsdens $et$ $al.$, 2002; Malinen $et$ $al.$, 2003; Matsuki $et$ $al.$, 2004). In addition, applications of 50-nuclease PCR for quantitative detection of $Listeria$ $monocytogenes$ and $Campylobacter$ $jejuni$ have been described (Novga $et$ $al.$,

2000b) as well as multiplex real-time PCR technology for detection and quantification of *Escherichia coli* O157:H7 (Ibekwe *et al.*, 2002).

Therefore, the real time (TaqMan) PCR has evolved into a fast yet sensitive, accurate tool for quantifying genes as marker bacteria in environmental samples (Huijsdens *et al.*, 2002; Harms *et al.*, 2003). The recent TaqMan MGB (Minor Groove Binder) probe with a No Fluorescent Quencher (NFQ) was developed to be more specific than traditional probes (Kutyavin *et al.*, 2000). For *Escherichia coli*, molecular detection procedures have been described based in the main on the *uidA* gene (Bej *et al.*, 1991a). Frahm and Obst (2003) proposed TaqMan PCR to detect *Enterococcus* spp. and *Escherichia coli*, but detection required a culture enrichment stage. Ideally, indicator bacteria should be dominant in feaces but absent in environmental samples, whereas *Escherichia coli*, although it is easily enumerated, is only subdominant in human faeces (Wilson and Blitchington, 1996; Suau *et al.*, 1999; Harmsen *et al.*, 2002; Hayashi *et al.*, 2002). Molecular methods enabled a free choice to be made of indicators without culture limitation. 16S rDNAs have been widely used in ecological and taxonomic species-specific studies (Amann *et al.*, 1995). The predominant faecal bacterial community was defined using culture-free approaches based on molecular techniques (Wilson and Blitchington, 1996; Suau *et al.*, 1999; Harmsen *et al.*, 2002; Hayashi *et al.*, 2002). A recent PCR detection technique (TaqMan) based on a 3V-Minor Groove Binder (MGB) probe was applied to the detection of faecal-dominant bacteria to assess feacal contamination in environmental samples (Rousselon *et al.*, 2004). Primers and probes used bacterial 16S ribosomal DNA (16S rDNA) as a gene marker and accurately defined with specificity a cluster of phylotypes within the Gram positive bacteria and low GC division. This cluster of phylotypes, called Fec1, corresponds to around 5% of human faecal microflora. Fec1 clustered 16S rDNA and strains (*Eubacterium rectale*) of faecal origin. A range of samples made up of feces and intestinal samples from humans and animals tested positive whereas other microbial ecosystems (soils, laboratory reactor, and subsurface water) were negative. In order to avoid problems related to DNA extraction efficiency, quantitative results took the form of the ratio between Fec1 16S rDNA and total bacterial 16S rDNA. The threshold of detection, defined as the ratio between Fec1 and total 16S rDNA, was measured as 0.006% (Rousselon *et al.*, 2004). Improved detection of pathogenic *Escherichia coli* (Ogunjimi and Choudary, 1999) by immuno-capture PCR, and the sensitive detection of *Salmonella* (Hoorfar *et al.*, 2000) and *Campylobacter* (Nogva *et al.*, 2000a) by real-time PCR have also been developed; but these procedures are all monospecific and are either difficult or very expensive for routine use in water testing laboratories. Moreover, the use of faecal *Bacteroidales* Q-PCR as a rapid method to complement traditional, culture-dependent, water quality indicators in systems where drinking water is supplied without chlorination or other forms of disinfection. A SYBR-green based, Q-PCR assay was developed to determine the concentration of faecal *Bacteroidales* 16S rRNA gene copies (Saundersa *et al.*, 2009).

## 3.4. MULTIPLEX REAL-TIME PCR ASSAY

The current real-time PCR assays for detecting such *V. cholerae* do not detect more than two gene targets in a single reaction (Lyon, 2001, Fukushima *et al.*, 2003). This limits the specificity of these described assays, as with the determination of an increased number of genome sequences, some gene targets used previously have now been shown that they are not unique to *Vibrio cholerae*. The availability of whole genome sequences of *Vibrio cholerae* (Heidelberg *et al.*, 2000) and of several closely related *Vibrio* spp. has also assisted in the identification of unique *Vibrio cholerae* gene sequences. A multiplex, real-time detection assay was developed targeting four genes characteristic of potentially toxigenic strains of *V. cholerae*, encoding: repeat in toxin (*rtxA*), extracellular secretory protein (*epsM*), mannose-sensitive pili (*mshA*) and the toxin coregulated pilus (*tcpA*). This assay was developed on the Cepheid Smart Cycler using SYBR Green I for detection and the products were differentiated based on melting temperature (Tm) analysis (Gubala, 2006). Validation of the assay was achieved by testing against a range of *Vibrio* and non-*Vibrio* species. The detection limit of this assay was determined to be $10^3$ CFU using cells from pure culture. This assay was also successful at detecting *Vibrio cholerae* directly from spiked environmental water samples in the order of $10^4$ CFU, except from sea water which inhibited the assay. The incorporation of a simple DNA purification step prior to the addition to the PCR increased the sensitivity 10 fold to $10^3$ CFU. This multiplex real-time PCR assay allows for a more reliable, rapid detection and identification of *Vibrio cholerae* which is considerably faster than current conventional detection assays (Gubala, 2006). Additionally, a real-time multiplex PCR assay was developed for the simultaneous detection of the thermolabile hemolysin (*tlh*), thermostable direct hemolysin (*tdh*), and thermostable-related hemolysin (*trh*) genes of *Vibrio parahaemolyticus*. The *tlh* gene is a speciesspecific marker, while the *tdh* and *trh* genes are pathogenicity markers (Nordstrom *et al.*, 2007).

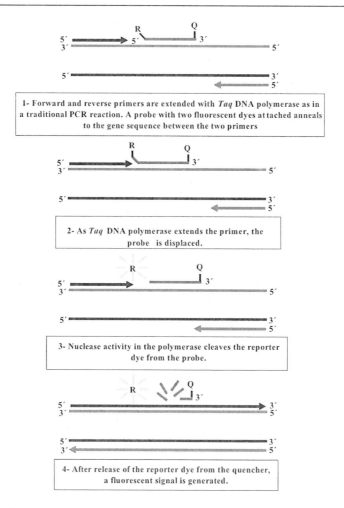

Figure 3. Schematic representation of Fluorogenic 5' nuclease in real-time PCR.

Moreover, a multiplex real-time PCR (R-PCR) assay was designed and evaluated on the ABI 7700 sequence detection system (TaqMan) to detect enterohemorrhagic *Escherichia coli* (EHEC) O157:H7 in complex samples. Three sets of primers and fluorogenic probes were used for amplification and real-time detection of a 106-bp region of the *eae* gene encoding EHEC O157:H7-specific intimin, and 150-bp and 200-bp segments of genes *stx1* and *stx2* encoding Shiga toxins 1 and 2, respectively (Sharma and Dean-Nystrom, 2003). A SYBR Green Light Cycler PCR assay using a single primer pair allowed simultaneous detection of *stx1* and/or *stx2* of *Escherichia coli* O157:H7 (Jothikumar and Griffiths, 2002). A multiplex real-time PCR assay was also developed for detection of *Legionella pneumophila* and *Legionella* spp. (Templeton *et al.*, 2003).

## 3.5. REVERSE TRANSCRIPTION POLYMERASE CHAIN REACTION (RT-PCR)

Reverse transcription polymerase chain reaction is a laboratory technique used for amplifying a defined piece of a ribonucleic acid (mRNA) molecule. The mRNA strand is first reverse transcribed into its DNA complement, followed by amplification of the resulting DNA using polymerase chain reaction (Figure 4). The two reactions are compatible enough that they can be run in the same mixture tube, with the initial heating step of PCR being used to inactivate the transcriptase enzyme (Saiki *et al*, 1988). Metabolic indicators such as the ability to carry out DNA synthesis have proved useful indicators of the viability of cells in the natural environment (Barer *et al.*; 1993, Kaprelyants *et al.*, 1993). Nucleic acids as viability indicators should have the additional advantage of specificity. In principle, the presence of one or another type of nucleic acid (DNA, rRNA, mRNA or tRNA) in bacterial cells might be a useful indicator of viability. Because of the presence of the 29-hydroxyl group of ribose, the phosphodiester bonds of RNA are more sensitive to hydrolysis than those of DNA, particularly in the presence of divalent cations. Therefore, RNA is more labile than DNA and more sensitive to degradation caused directly by deleterious treatments, such as heating or acidification. Therefore, nucleic acid-based methods that could be applied directly to samples to give an indication of the viability of any microorganisms present should be of massive significance for food, industrial, environmental, and medical applications. mRNA is broken down rapidly in living bacterial cells, with most mRNA species having a half-life of only a few minutes (Alifano *et al.*, 1994; Belasco, 1993). Consequently, detection of mRNA might be a good indicator of living cells or those only lately dead at the time of sampling. Detection of mRNA by Northern blot hybridization has been used as an indicator of microbial metabolic activity in aquatic and soil samples (Tsai and Olson; 1990; Pichard and Paul, 1993; Jeffrey *et al.*, 1994), and detection of mRNA by reverse transcription-PCR (RT-PCR) was used to monitor gene expression in activated sludge (Selvaratnam *et al.*, 1995). Whereas, these studies definitely indicated the activities of viable cells in natural environments, their main purpose was not to differentiate between living and dead cells. Few studies have exclusively investigated the relationship between detection of microbial mRNA and viability. RT-PCR was used to examine *Legionella pneumophila* and *Vibrio cholerae* exposed to heat or starvation, respectively, and detected specific mRNA only in samples that contained viable cells detected by culturing (Bej *et al.*, 1991c; Bej *et al.*, 1996). Similarly, Patel *et al.* (1993) successfully assessed the viability of heat-killed *Mycobacterium leprae*, detecting a heat shock protein mRNA in living cells. Recently, RT-PCR method for the specific detection of viable *Listeria monocytogenes* was developed (Klein and Juneja, 1997) based on the detection of mRNA from *iap*, *hly*, or *prfA* genes. Despite their potential advantages, mRNA-based methods have demonstrated difficult to use because of the complexity of the protocols, the practical problems of extracting detectable levels of intact mRNA from small numbers of bacteria, and a lack of basic information about the significance of detecting mRNA in stressed cells. Recently, approaches were developed for detecting specific mRNA from *Escherichia coli* and to examine the relationship between viability and the presence of mRNA (Sheridan *et al.*, 1998). Since any relationship between mRNA and viability may depend on the method used to inactivate cells or the type of mRNA required, another study exposed the cells to two different stress treatments (heat and ethanol) and assayed mRNA from three different genes (*rpoH*, *groEL*, and *tufA*) (Sheridan *et al.*, 1998). These genes were chosen as representing a gene encoding an abundant cellular housekeeping protein (*tufA*) (Bosch *et al.*, 1983; Van der Meide *et al.*, 1983) and genes associated with a stress response regulon (*rpoH* and *groEL*) (Erickson *et al.*, 1987; Bakau, 1993; Hendrick and Hartl, 1993). Of the different types of nucleic acid, mRNA is the most promising candidate for an indicator of viability in bacteria, but its persistence in dead cells depends on the inactivating treatment and subsequent holding conditions (Sheridan *et al.*, 1998). This may be correlated with the stability of the different mRNA sequences targeted by the oligonucleotide primers designed (Table 2) or to the relative abundance of each mRNA type in a given cell. Either way, it appears that the type of mRNA selected for detection of viable bacteria will be important. Furthermore, rRNA has been suggested as an indicator of viability in *Mycobacterium smegmatis* (Van der Vliet *et al.*, 1994). A correlation between viability and cellular rRNA content was also observed in *Escherichia coli* cells undergoing starvation (Davis *et al.*, 1986) but did not occur in starved *Azotobacter agilis* cells (Sobek *et al.*, 1966). However, 16S rRNA from heat- or ethanol-killed *Escherichia coli* cells did not disappear after 16 hours of incubation in broth at room temperature. It was not, therefore, a useful indicator of viability for the time frame over which our experiments were conducted (i.e., 16 hours) (Sheridan *et al.*, 1998). There is very little information about the persistence of tRNA in injured or dead cells. tRNA survived much longer than rRNA in cells of *Escherichia coli* that had died of starvation, and it is therefore unlikely to be a good indicator of viability (Davis *et al.*, 1986). A good correlation was showed between the presence of mRNA and the viability of *Listeria monocytogenes* when comparing growing cells with those killed by autoclaving (Klein and Juneja, 1997). Methods existed for detecting mRNA in single bacterial cells (Hodson *et al.*, 1995), and further improvements in technology are likely to increase the application of these methods in microbial monitoring and ecology.

## 3.6. DNA FINGERPRINTING TECHNIQUE

DNA fingerprint means a pattern generated during analysis of DNA that is unique to an individual organism. DNA fingerprinting is a laboratory procedure that requires as follows: Isolation of DNA, DNA must be recovered from the microorganisms, only a small amount is needed; Cutting, sizing, and sorting the DNA, restriction enzymes are used to cut the DNA at specific places and the DNA pieces are sorted according to size by electrophoresis through agarose gel electrophoresis; transfer of DNA to nylon, The DNA pieces are transferred to a nylon sheet by placing the agarose gel and nylon next to each other overnight; probing, the DNA fingerprint is generated by adding tagged probes to the nylon sheet, each probe typically sticks in only one or two specific places, wherever the sequences match; and generating DNA fingerprint, the final DNA fingerprint is built from several different probes and resembles as the bar codes used at the grocery counter.

Fingerprinting techniques such as enterobacterial repetitive intergenic consensus polymerase chain reaction (ERIC-PCR) and amplified fragment length polymorphism (AFLP) have been used to identify the differences in host-specific *Escherichia coli* strains regarding their pathogenicity (Parveen *et al.*, 1999; Guan *et al.*, 2002; Leung *et al.*, 2004). Rep-PCR is a DNA fingerprint technique that uses repetitive intergenic DNA sequences to differentiate between sources of faecal pollution (Dombek *et al.*, 2000). With this technique, DNA between adjacent repetitive extragenic elements is amplified using PCR to produce various size DNA fragments (Farber, 1996; Dombek *et al.*, 2000). The PCR products are then size-fractionated by agarose-gel electrophoresis to produce specific DNA fingerprint patterns. These fingerprint patterns can then be analyzed using pattern recognition computer software (Dombek *et al.*, 2000). In this respect, the repetitive sequence-based polymerase chain reaction (Rep-PCR) has been successfully utilized to classify the number of different bacterial species, which has been difficult to do based on the standard culture-based methods (Dombek *et al.*, 2000; Seurinck *et al.*, 2003). Although the PCR-based techniques have been widely applied for direct pathogen detection, some problematic issues still exist. Thus, the rep-PCR technique was chosen because this technique is simple, can differentiate between closely related strains of bacteria, and can be used for high-throughput studies (Versalovic *et al.*, 1994). Previously, Rep-PCR has been used successfully to classify and differentiate among strains of *Escherichia coli* (Lipman *et al.*, 1995). Dombek *et al.* (2000) used Rep-PCR with BOX A1R primer to discriminate between human and six species of animal (cows, pigs, sheep, chickens, geese and ducks) feacal pollution. Overall, they suggested that Rep-PCR is a useful and effective method for rapidly classifying and grouping *Escherichia coli* isolates from humans and animals. Recent study achieved by Carson *et al.* (2003) was carried out a comparison study of ribotyping and Rep-PCR on eight host classes (human, cattle, pig, horse, dog, chicken, turkey, and goose) and found that the average rate of correct classification for ribotyping was 73 versus 88% for Rep-PCR. It was concluded that Rep-PCR was more reproducible, accurate, and efficient than ribotyping. However, in this comparison study, it was only used one restriction enzyme (*Hind*III) for ribotyping versus two as recommend by Samadpour (2002). Holloway (2001) using the protocol of Dombek *et al.* (2000), performed Rep-PCR to determine animal host type for ninety one *Escherichia coli* and sixty eight *Enterococcus faecalis* strains from human, swine, cattle and poultry faeces. Contrary to the results from Carson *et al.* (2003) and Dombek *et al.* (2000), Holloway (2001) did not observe any significant clustering of *Escherichia coli* or *Enterococcus faecalis* strains by animal type. Holloway (2001) suggested that too few strains may have been tested in his study. In conclusion, Holloway (2001) stated that this technique is not ready and reliable for the identification of the source of faecal contamination in water and that a large sample size may be necessary for this approach.

## 3.7. T-RFLP AND LH-PCR TECHNIQUES

Terminal restriction fragment length polymorphism (T-RFLP) allows the fingerprinting of a community by analyzing the polymorphism of a certain gene. It is a high-throughput, reproducible method that allows the semi-quantitative analysis of the diversity of a particular gene in a community. The Figure 5 demonstrates the procedure and the basis of the method. The DNA is harvested from the analyzed sample. The gene of interest is amplified using the polymerase chain reaction (PCR) with a fluorescently labeled primer. This yields a mixture of amplicons of the same or similar sizes with a fluorescent label at one end. After purification, the amplicon mixture is digested with a restriction enzyme, which generates fragments of different sizes (A-F). These are separated through gel or capillary electrophoresis. A laser reader detects the labeled fragments and generates a profile based on fragment lengths.

While as, length heterogeneity analysis of PCR amplified genes (LH-PCR) utilizes naturally occurring differences in the lengths of amplified gene fragments. The abbreviation LH-PCR has been used so far only for the analysis of 16S rRNA fragments. Generally, strong heterogeneity seems to be most common in the apical helices of ribosomal molecules, i.e. those ending in a hairpin loop. The sizes of the fragments on the polyacrylamide gel after electrophoresis can be compared against 16S rRNA gene databases to specify microbial groups that may correspond in size to the size of the fragments. Length heterogeneity analysis of the ribosomal intergenic spacer region (RISA) is analogous to LH-PCR. RISA involves the PCR amplification of the region of the rRNA gene operon between the small (16S RNA) and large (23S RNA) subunits called the intergenic spacer (IS) region (Figure 6). By using oligonucleotide primers targeted to conserved regions in the 16S and 23S genes, RISA fragments can be generated from the members of the microbial community. The fragments are separated using polyacrylamide gel electrophoresis and detected with fluorescent DNA dyes or PCR primers. The length variation of the spacer region is considerable (150-1500 bp) between different species. Both LH-PCR and RISA can be automated with a sequencing machine. The main limitation of RISA is the unpredictability and random pattern of variations in spacer size. In the case of multiple ribosomal operons within a strain, the sizes and sequences of different 16S-23S RNA PCR amplicons may vary considerably in a single bacterial cell. In community profiling this means that a single species may contribute more than one peak or band to the community profile. LH-PCR separates PCR products for host-specific genetic markers based on length of amplicons (Bernhard and Field, 2000a). LH-PCR can quickly provide a profile of amplicon variety in complex mixtures of PCR products (Suzuki et al., 1998).

T-RFLP and LH-PCR take advantages of unique genetic markers to distinguish between different microorganisms that allows identifying the origin of faecal pollution (Bernhard and Field, 2000a, b; Blackwood *et al.*, 2003; Nagashima *et al.*, 2003). Recent advances in molecular characterization of *Cryptosporidium* spp. achieved sensitive identification and differentiation of this pathogen in water samples for the evaluation of the faecal contamination source (Xiao *et al.*, 2001). The host specific 16S ribosomal DNA (rDNA) genetic markers technique discriminates members of mixtures of bacterial gene sequences by detecting differences in the number of base pairs in a particular gene fragment (Bernhard and Field, 2000a & b). T-RFLP uses restriction enzymes on PCR amplicons to determine unique size fragments among fluorescently labeled terminal end fragments (Bernhard and Field, 2000a). LH-PCR and T-RFLP analyze differences in the lengths of gene fragments due to insertions and deletions to estimate the relative abundance of each fragment (Bernhard and Field, 2000a). This method helps to decrease some of the problems associated with the under sampling of diversity in a microbial community and the uncertainty of bias due to the reannealing kinetics in the cloning process by PCR (Suzuki et al., 1998). 16S rDNA markers was developed, based on faecal anaerobes (*Bacteroides* and *Bifidobacterium*), to distinguish human and cow faecal pollution (Bernhard and Field, 2000a). Strict anaerobes were chosen because they are restricted to warm-blooded animals, make up a large portion of the feacal bacteria, and do not survive long once deposited in waters. *Bacteroides* and *Bifidobacterium* have had limited use as indicators of faecal pollution because they are difficult to grow in culture media. The use of molecular methods versus culture-based methods improved the ability for their use in water quality monitoring. The *Bacteroides–Prevotella* group was a better indicator than the *Bifidobacterium* species due to the ease of detection and longer survival in water (Bernhard and Field, 2000a). Bernhard and Field (2000b) also tested their approach on faeces from human, sewage and cattle sources and found their method was successful in being able to distinguish sources. Since only human and cattle markers were studied, further research needs to be conducted on other sources of faecal contamination such as wildlife and domestic animals other than cattle.

## 3.8. RIBOTYPING TECHNIQUE

Ribotyping is also referred to a DNA fingerprint technique and can detect, identify and classify bacteria based upon differences in rRNA. It generates a highly reproducible and precise fingerprint that can be used to classify bacteria from the genus through and beyond the species level. This method is based on that DNA is extracted from a colony of bacteria and then restricted into discrete-sized fragments. The DNA is then transferred to a membrane and probed with a region of the rRNA operon to reveal the pattern of rRNA genes. This pattern is recorded, digitized and stored in a database. It is variations that exist among bacteria in both the position and intensity of rRNA bands that can be used for their classification and identification. So, ribotyping is a method of identifying microorganisms from the analysis of DNA fragments produced from restriction enzyme digestion of genes encoding their 16S rRNA (Farber, 1996; Aarnisalo *et al.*, 1999; Samadpour, 2002). The ribotyping protocol provides a DNA fingerprint of bacterial genes coding for ribosomal ribonucleic acids (rRNA), which are highly conserved in microorganisms (Farber, 1996; Samadpour, 2002). Unique strains of *Escherichia coli* are adapted to their own specific environment (intestines of host species), and as a result differ from other strains found in other host species. The ribotyping technique is also used for microbial source tracking. To use the microbial source tracking method, collections of potential source material (faecal samples of all potential sources in the watershed) must be collected and sub-typed. The genetic fingerprints of the bacterial isolates from the water samples can then be compared to those of the bacteria from the suspected animal sources (Samadpour, 2002). Ribotyping does not contain sequencing, it alternatively measures the unique pattern generated when DNA from a specific organism is subject to restriction enzyme digestion and the fragments are separated and probed with a ribosomal RNA probe (Farber, 1996; Samadpour, 2002).

The ribotyping procedure is also performed as follows: bulk DNA or DNA encoding 16S rRNA and related genes within the rRNA operon is polymerase chain reaction (PCR) amplified, treated with one or more restriction enzymes, separated by gel electrophoresis and then probed. The pattern is generated from the DNA fragments on the gel is digitized and a computer is used to make comparisons of the patterns with patterns from reference *Escherichia coli* in a database. The unique pattern of DNA bands (riboprints), are used to determine the host of the original bacteria (Parveen *et al.*, 1999; Carson *et al.*, 2001; Farag *et al.*, 2001; Samadpour, 2002). Ribotyping is both a rapid and specific method of bacterial identification. The microbial source tracking methodology for ribotyping has been under development for the past 12 years, and it has been applied to over 80 studies in the US and Canada (Samadpour, 2002). The choice in restriction enzymes used for ribotyping is critical, and that double enzyme analysis should be used to identify clones to a higher degree of accuracy, as single enzyme digestion is insufficient (Samadpour, 2002). Carson *et al*. (2001) and Parveen *et al*. (1999) have tested the average rate of correct classification (ARCC) achieved by ribotyping when differentiating between human and non-human sources of feacal pollution. Carson *et al*. (2001) found that when using discriminant analysis the rate of correct classification from each of eight known sources (human, cattle, pig, horse, dog, chicken, turkey, and goose) ranged between 49 and 96 %. They found a higher classification accuracy when the analysis was limited to three host sources (i.e. cattle, pigs and humans), and were able to achieve a 97 % ARCC by grouping the non-human riboprints and comparing them to human riboprints. Carson *et al*. (2001) used only one restriction enzyme (*Hind*III) in their study. Parveen *et al*. (1999) used a number of restriction enzymes including *Hind*III, *Eco*RI, *Sal*I and *Bgl*I for discriminant analysis of ribotype profiles to correctly classify 97 and 67 % of the non-human and human source isolates using the ribotype method and their ARCC was 82%.

## 3.9. FLUORESCENCE *IN SITU* HYBRIDIZATION (FISH)

The term *in situ* hybridization is restricted to whole cell hybridizations wherein the cells are detected in their natural microhabitat. Fluorescent *in situ* hybridization can be used to detect, quantify and identify specific microorganisms or bacterial community while keeping their morphological integrity, without nucleic acid extraction. Sample cells are fixed with chemicals to increase their permeability and allow the probe to enter the cells. Fixation chemicals are precipitants like ethanol or methanol and cross-linking agents like aldehydes such as paraformaldehyde. After fixation the sample is spotted into wells of Teflon-covered slides, which have been gelatin-coated to aid the attachment of the cells. The sample is air-dried and then dehydrated by serial immersion of the slide in ethanol series with increasing concentration. After hybridization with fluorescence labeled oligonucleotide probe (Figure 7) in hybridization buffer, the excess probe is washed, and the slides air dried and then visualized using fluorescence microscopy. Whole cell hybridization with fluorescently labelled probes can also be combined with flow cytometry for rapid counting or collection of cells. Most of the nucleic acid protocols use molecular hybridization properties, which include the complementary sequence recognition between a nucleic probe and a nucleic target. A hybridization reaction can be achieved between a nucleic DNA probe and a chromosomic DNA sequence (DNA–DNA hybridization) or an rRNA or tRNA sequence (DNA–RNA hybridization). Specificity, here, depends on the phylogenetic degree of conservation of the target within the taxonomic target group. These methods provide taxonomic information at different levels, such as classes, genera, species or subspecies. Some of them can be performed without the need for a complex cultivation step, thus permitting the detection of specific bacteria within hours, instead of few days required with the cultivation-based methods.

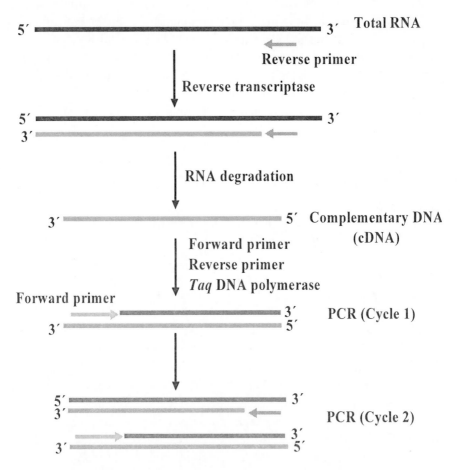

Figure 4. Schematic demonstration of reverse transcription –PCR (RT-PCR).

Conventionally, Fluorescence *in situ* hybridization methods are based on the use of conventional DNA oligonucleotide probes, containing around 20 bases. Hybridization must be carried out under high stringent conditions for proper annealing of the probe to give the required affinity and specificity for binding. Stringency can be adjusted by varying the formamide concentration and the hybridization temperature, but nucleic acid multiplexing is more difficult because different probes differ in their stringency requirements (Moter and Gobel, 2000). Whole cell *in situ* hybridization with fluorescently labeled oligonucleotide probes has been widely introduced to environmental microbial ecology (Head *et al.*, 1998; Okabe *et al.*, 1999). Moreover, the value of the technique is often not appreciated by the choice of binding site on the target rRNA. Fluorescence in situ hybridization is also based on the specific binding of nucleic acid probes to specific regions on rRNA. The use of rRNA-targeted oligonucleotide probes with detectable marker molecules enables the visualization of whole cells and the identification and study of microbes in situ. There are several reasons why rRNA is particularly suitable for hybridization. The number of rRNA molecules per cell is normally more than 1000 and the probe design is relatively simple, due to the sequence collections available in electronic databases, especially for the 16S rRNA (Amann *et al.*, 1997). In *Escherichia coli*, the target site of the DNA probe strongly affects the signal intensity of the fluorescently labeled oligonucleotide probe (Fuchs *et al.*, 1998). Fuchs *et al.* (1998) classified the site accessibility and binding affinities of DNA oligomeric probes to 16S rRNA to 6 classes (Class I being the highest affinity and producing the brightest fluorescence) according to the 16S rRNA target site of *Escherichia coli*. Therefore with traditional DNA oligonucleotide probes, the binding affinity and resulting fluorescent intensity would be remarkably lower for Class II–VI domains than it would be in the Class I domain.

Peptide nucleic acid (PNA) molecules are DNA mimics, where the negatively charged sugar–phosphate backbone is replaced by an achiral, neutral polyamide backbone formed by repetitive units of N- (2-aminoethyl) glycine. PNA can hybridize to complementary nucleic acid targets obeying the Watson–Crick base pairing rules (Egholm *et al.*, 1993). PNA molecules have unique hybridization characteristics, exhibiting rapid and stronger binding to complementary targets and a lack of electrostatic repulsion. Also, the unnatural PNA backbone is not degraded by enzymes such as nucleases and proteases (Demidov *et al.*, 1994).

**1- Extraction of Total DNA from the bacterial community**

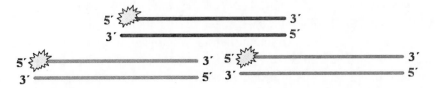

**2- PCR with a fluorescent labeled forward primer of gene interest (16 S rRNA)**

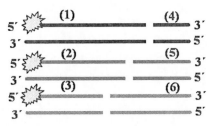

**3- Digestion of PCR products with restriction enzymes**

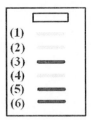

**5- Separation of fragments in polyacrylamide gel electrophoresis**

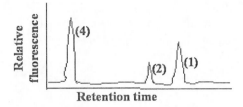

**5- Identification based on retention time**

Figure 5. Schematic presentation of terminal restriction fragment length polymorphism (T-RFLP).

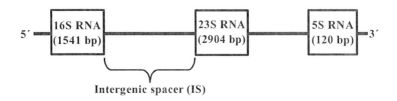

Figure 6. Schematic diagram showing location of intergenic spacer (IS).

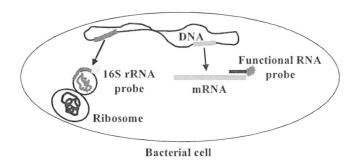

Figure 7. In situ hybridization of the functional and taxon specific probes within bacterial cell.

Recent methods have been developed using fluorescently labeled PNA FISH probes targeting specific rRNA sequences of Gram-negative and Gram-positive bacteria for rapid confirmation of *Escherichia coli, Pseudomonas aeruginosa, Staphylococcus aureus, Salmonella,* and *Mycobacterium tuberculosis* from agar colonies, and identification of *Helicobacter pylori* in environmental samples (Stender *et al.*, 1999; Perry-O Keefe *et al.*, 2001; Azevedo *et al.*, 2003). Typically the probes used for PNA FISH are shorter (optimum 15 bases), thus have a greater permeability through the bacterial cell wall and require a simpler fixation step than DNA oligonucleotides (Stender *et al.*, 2002). Moreover, the use of a low salt buffer for PNA hybridization decreases the stability of the secondary structures of rRNA, thus allowing PNA probes to hybridize to less accessible targets (Perry-O Keefe *et al.*, 2001; Armitage, 2003).

*Campylobacter jejuni* and *Campylobacter lari* are found to be an important cause of drinking water related epidemics (Broczyk *et al.*, 1987; Hanninen *et al.*, 2003). The most important species causing bacterial gastroenteritis are *Campylobacter jejuni* and its close relative *Campylobacter coli* (Skirrow, 1990; Wassenaar and Newell, 2000). Several molecular methods for identifying *Campylobacter* spp. have been developed (Wassenaar and Newell, 2000; Steinhauserova *et al.*, 2001; Keramas *et al.*, 2003). These methods are typically based on the analysis of PCR products and provide species specific detection of *Campylobacteria*, but they are not easy and have not yet been validated for the use on different viability states of bacteria. A FISH method for detection of thermotolerant *Campylobacter* spp. using conventional DNA probes has been published (Buswell *et al.*, 1998; Moreno *et al.*, 2001 & 2003). Recent report demonstrated a rapid PNA FISH method that can identify clinically important thermotolerant *Campylobacter* spp. using the PNA chemistry to target a low DNA affinity binding site (Lehtola *et al.*, 2005). This study added advantage that is the ability of PNA sequences to target regions of the 16S rRNA that conventional DNA sequences would have lower accessibility. This attribute of PNA offers the potential to target more potential binding sites on the rRNA, facilitating the design of species-specific probes. Based on designed probes, binding complimentary rRNA sequences, this method allows to detect metabolically active microbial community of interest and also to quantify them (Pernthaler and Amann, 2004; Savichtcheva *et al.*, 2005). This technique was applied for rapid detection, identification, and enumeration of *Escherichia coli* cells in municipal wastewaters (Stender *et al.*, 2001).

The advantage of *in situ* hybridization is that it can give quantitative information about the number of specific microorganisms. However, *in situ* hybridization also has limitations. Frequently encountered problems include no signals or low signal intensity. This can be caused by noncomplementarity of probe and target, ineffective probe labeling or nonoptimal hybrization conditions. Low signal intensity can be caused by small numbers or insufficient accessibility of the target molecules (rRNA). Dormant or metabolically inactive cells may not contain enough rRNA to show fluorescent signals after hybridization. The factors that delay probe binding include limited diffusion of rRNA-targeted probes due to cell boundaries (probably cell wall) and prevention caused by higher order structures in the ribosomes. However, quantification of the number of microorganisms with *in situ* hybridization can be problematic when cells have irregular shapes or when they form chains or compact micro-colonies. *In situ* delectability is also influenced by non-specific binding and autofluorescence of cells and surrounding material.

## 3.10. DNA MICROARRAY-BASED TECHNOLOGY

Several nucleic acid-based methods have been developed for the rapid detection of pathogens in food, soil, and water with high degrees of sensitivity and specificity and without the need for complex cultivation (Bej *et al.*, 1991b; Wilson *et al.*, 2002; Bej, 2003; Call *et al.*, 2003; Lee *et al.*, 2003). In general, these methods allow detection within hours, rather than days as is normally required by culture techniques. Due to its high sensitivity and specificity, PCR is the most commonly employed molecular tool (Foy *et al.*, 2001). A major limitation to this approach is the utilization of one specific primer pair per gene detection reaction. Although multiple primer sets may be successfully combined in one reaction, they rarely exceed more than six primer sets due to the generation of nonspecific products or false negatives. Another difficulty with multiplex PCR is that it requires additional postamplification analysis to differentiate the products. Size separation by electrophoresis is frequently used to discriminate multiplex PCR products, but this requires additional work and that the amplicons of each reaction be significantly different in size, which can limit the primer pairs that can potentially be multiplexed. Consequently, general pathogen detection by PCR can be both work-intensive and costly.

Genome sequencing projects have stimulated fundamental changes in experimental methodology from those focused on "one gene at a time" to those aimed to study thousands of genes or proteins at once. Thus, DNA microarray-based technologies have revolutionized the ability to carry out hundreds or thousands of hybridization reactions at a time, showing huge screening capacity resulting in a high level of sensitivity, specificity, and output capacity (Vora *et al.*, 2004). A DNA (DNA chip, gene chip or biochip) is a collection of DNA probes attached to a solid surface, such as glass chip forming an array. Sample DNA or RNA is extracted, RNA is reverse transcribed to cDNA and the DNA or cDNA is labelled with fluorescent labels. The labelled DNA is denatured and hybridized with the probes on the array. Unbound probes are washed away and the array is visualized using confocal laser microscope scanner (Figure 8). Microarray is fast and cost effective tool to detect specific microorganisms, since thousands of hybridizations can be performed on one chip. The results are well comparable, but the optimization of hybridization conditions for a large number of probes may be challenging. Also the designing of probes and arrays is very time consuming. Microarrays represent an important advance in molecular detection technology, allowing the simultaneous detection of specifically labeled DNAs from many different pathogenic organisms on a small glass slide containing thousands of surface immobilized DNA probes. Both basic types of microarrays, i.e., immobilized oligonucleotide probes and PCR amplicons, have been used successfully to detect (Wilson *et al.*, 2002) and/or characterize (Bekal *et al.*, 2003) pathogens.

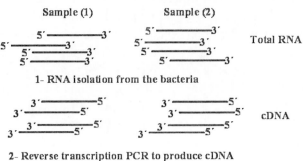

1- RNA isolation from the bacteria

2- Reverse transcription PCR to produce cDNA

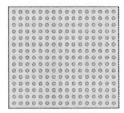

3- Labeling of cDNA with two fluorescent dyes

4- Hybridization with unlabelled probes on DNA chip

5- Visualization by confocal laser microscope scanner

Figure 8. Representation of microarray analysis of RNA isolated from two samples using two different labels.

The DNA microarray technique has been successfully applied for detection of microbial virulence factors (Chizhikov *et al.*, 2001; Vora *et al.*, 2004) and for quantification of bacterial DNAs in environmental samples (Cho and Tiedji, 2002). Genetic characterization of *Salmonella enterica* strains (Alvarez *et al.*, 2003) and detection of different pathogens including enterohemorrhagic *Escherichia coli* (EHEC) O157:H7 (Vora *et al.*, 2004), six species of *Listeria* genus (Volokhov *et al.*, 2002), and human group A rotaviruses (Chizhikov *et al.*, 2002) have also been performed using the microarray analyses. As the sensitivity of microarrays hybridized with total genomic DNA from complex mixtures is usually insufficient to provide detection of low pathogen concentrations (Rhee *et al.*, 2004), the hybridized DNA (target) usually consists of PCR amplicons (Call *et al.*, 2001; Wilson *et al.*, 2002). This mode of pathogen detection requires the combination of many PCRs prior to their hybridization on microarrays. Previous study (Wilson *et al.*, 2002) used 140 amplicons to characterize eighteen pathogenic species, therefore limiting the use of microarrays for routine detection of pathogens in wastewater. Target DNA amplification with universal primers to ubiquitous genes prior to microarray hybridization can avoid this limitation (Cook *et al.*, 2004; Keramas *et al.*, 2003; McCabe *et al.*, 1999; van der Giessen *et al.*, 1992). The *cpn60* gene codes for GroEL, an essential, highly conserved chaperonin protein which displays moderate DNA sequence diversity, making this gene useful in bacterial taxonomy applications (Goh *et al.*, 1996; Broussea *et al.*, 2001; Hill *et al.*, 2002). However, within the Enterobacteriaceae, 16S rRNA and *cpn60* sequences may share sufficient similarity to generate cross-hybridization reactions, even when short oligonucleotides are used as probes. As the majority of water pathogens belong to this family, differentiation on the basis of 16S rRNA and *cpn60* sequences is challenging. However, sequence diversity within the *wecE* gene, an Enterobacteriaceae-specific gene which forms part of the *wec* gene cluster involved in enterobacterial common antigen biosynthesis, has been shown to discriminate among the Enterobacteriaceae most frequently found in water (Bayardelle and Zafarullah, 2002).

In recent study, a specific and sensitive microarray was designated to be used for the detection of several bacterial species in wastewater samples. An oligonucleotide-based microarray designed with probes specific for the universal targets of 16S rRNA and *cpn60* genes in addition to the Enterobacteriaceae-specific *wecE* genes of several pathogens usually encountered in wastewater. Amplification and fluorescent labeling of the 16S rRNA, *cpn60*, and *wecE* genes from extracted community DNA show specific detection of each of the microorganisms studied when hybridized to oligonucleotide probes printed on the wastewater prototype microarray. Although sensitivity may vary depending on the microorganisms tested, detection sensitivity can be increased by targeting amplicons specific for a limited group of bacteria instead of universal taxonomic amplicons from a broad spectrum of bacteria. This study demonstrated the capability of using DNA microarrays in the detection of waterborne pathogens within mixed populations (Maynard *et al.*, 2005). Moreover, recent study demonstrated that the use of microarray-based methods, which offer the ability to simultaneously interrogate for the presence of several genes, are ideally suited for the determination of genotype, and hence, biological potential. A 90-plex PCR assay was designed and combined with a long oligonucleotide DNA microarray that could simultaneously identify diagnostic regions specific for species, serogroup, biotype, antimicrobial resistance, and pathogenicity markers for *Vibrio cholerae*, *Vibrio parahaemolyticus*, *Vibrio vulnificus*, and *Vibrio mimicus* (Vora *et al.*, 2005). This method was validated and demonstrated its utility in accurately identifying human pathogenic strains, as an examination tool for monitoring genetic heterogeneity, and as a method capable of detecting VBNC cells. In addition to accurate identification, the microarray simultaneously provided evidence of antibiotic resistance genes and mobile genetic elements, such as sulfamethoxazole-trimethoprim constins and class I integrons, and common toxin (*ctxAB*, *rtxA*, *hap*, *hlyA*, *tl*, *tdh*, *trh*, *vvhA*, *vlly*, and *vmhA*) and pathogenicity (*tcpA*, type III secretion system) genes that are associated with pathogenic *Vibrio* (Vora *et al.*, 2005). The versatility of this method was further underscored by its ability to detect the expression of known toxin and virulence genes from potentially harmful viable but nonculturable organisms. This molecular identification method provides rapid and definitive information that would be of value in epidemiological, environmental, and health risk assessment surveillance. This study has outlined the development of a molecular detection technology that addresses the constraints characteristic of conventional microbiological detection and identification methods. In contrast to other molecular typing methods often used to characterize pathogenic *Vibrio spp.*, this method directly and simultaneously characterizes genetic markers valuable for the accurate detection of pandemic isolates, disease causation, and potential treatment-based resolution. The microarray-based method (and further elaborations) can provide a valuable tool for the identification of genetic assemblages associated with particular types of infection, VBNC bacteria-contaminated samples, epidemiological tracking, and environmental investigation efforts.

# Chapter 4

# ASSESSMENT OF LIVE OR DEAD BACTERIAL PATHOGENS

Molecular detection methods based on nucleic acids are able to detect VBNC bacteria to a certain extent (Huq *et al.*, 2000). On the opposite, detection of nucleic acids of a bacterium does not tell much about its viability, a key factor for its pathogenicity. Treatment procedures of drinking water attempt to kill pathogenic bacteria. Therefore, the physiological state of bacteria in drinking water is a major issue for drinking water safety (Hoefel *et al.*, 2005). In the past decade, a broad spectrum of fluorescent dyes was developed to assess the physiological state of bacteria at the single cell level. An excellent overview on the characteristics of these dyes is given by Joux and Lebaron (2000). The dyes cover a broad spectrum of bacterial features ranging from dehydrogenase activity to membrane potential and integrity. With a set of these dyes the effect of treatment and disinfection procedures on bacterial cultures could be observed and the recovery potential on culture media was assessed (Nebe-von-Caron *et al.*, 2000; Berney *et al.*, 2006). These studies showed that, during disinfection, metabolic functions decreased first, followed by decreasing membrane potential. At last, the loss of membrane integrity was usually observed and associated with the loss of recovery of the bacteria on media and interpreted as cell death. The question, if bacteria possibly can recover before or when they enter a human host, is a very crucial one. Though, the loss of membrane integrity is now usually interpreted as cell death, it was numerously shown that membrane damages can be recovered (Chilton *et al.*, 2001). Therefore, many authors (Joux and Lebaron, 2000; Berney *et al.*, 2006) recommend to use more than one or a whole set of physiological stains if crucial disinfection processes have to be followed, and to be aware of the restrictions of a potential recovery that can be strongly influenced by the conditions before and after the treatment.

The use of DNA-based molecular detection tools for bacterial analysis is disadvantaged by the inability to distinguish signals deriving from live and dead cells. The detection of live cells is typically most applicable in molecular detections. The selective detection of live microorganisms creates a great challenge for molecular DNA-based diagnostic technology due to the persistence of DNA in the environment after cell death. Often, significant data can only be obtained if dead cells are excluded from the analysis. For bacteria, a relatively recent approach is the treatment of samples with the DNA-intercalating dyes ethidium monoazide (EMA) or propidium monoazide (PMA) (Nogva *et al.*, 2003; Rudi *et al.*, 2005a & b; Nocker *et al.*, 2006). This fast and simple treatment was proposed to lead to the exclusion of membrane-compromised bacterial cells from the analysis. Live cells are traditionally considered to have the potential for an active metabolism and to be intact, whereas lack of membrane-integrity has been accepted as an indication of cell death. While loss of membrane integrity is a very conservative definition of death and dead cells, after having lost viability, can still have intact membranes for some time, the exclusion of damaged cells would pose an important step in limiting DNA-based detection to a more relevant faction of a bacterial population. In addition to bacterial diagnostics, the PMA method has also been successfully used for assessing fungal viability by differentiating between live and heat-inactivated cells of different infectious fungi (Vesper *et al.*, 2008). The principle is based on the hypothesis that PMA (or EMA), when added to a mixture of intact and membrane-compromised cells, selectively enters only the compromised cells. Once inside the cell, it intercalates into nucleic acids. The presence of an azide group permits crosslinking the dye to the DNA by exposure to strong visible light. The light leads to the formation of a highly reactive nitrene radical, which can react with any organic molecule in its proximity including the bound DNA. This modification strongly inhibits the PCR amplification of the extracted DNA. At the same time when the crosslinking occurs, the light reacts unbound excess dye with water molecules. The resulting hydroxylamine is no longer reactive, so the DNA from cells with intact membranes is not modified in the DNA extraction procedure. While the mechanism of action of these chemicals has not been explained yet and could be result of a combination of factors, the overall result of treatment of membrane-compromised cells is a reduction in PCR amplification. Comparing the two dyes, PMA has the advantage over EMA in being more selective for dead cells as it is more membrane-impermeant (Nocker *et al.*, 2006; Flekna *et al.*, 2007; Cawthorn and Witthuhn, 2008) and it has not been observed to enter intact cells of many bacterial species after 5 min of exposure. The use of PMA for live-dead distinction has so far been validated in combination with quantitative PCR and end-point PCR-based denaturing gradient gel electrophoresis (DGGE) (Nocker *et al.*, 2007a & b).

Very recent study (Nocker *et al.*, 2009) evaluated the suitability of propidium monoazide (PMA) treatment to exclude isopropanol-killed cells from detection in defined mixtures using diagnostic microarray technology. The organisms comprised *Pseudomonas aeruginosa*, *Listeria monocytogenes*, *Salmonella typhimurium*, *Serratia marcescens*, and *Escherichia coli* O157:H7. This study has for the first time combined PMA treatment of mixtures of live and dead bacterial cells with microarray analysis of PCR products. The effect of sample treatment on microarray signals correlated very well with the results from Q-PCR analysis. Although microarray detection has a lower intrinsic sensitivity, its benefits lie in its high-throughput and its versatility. The highly multiplexed format of the technology allows for rapid detection of specific gene targets of a great number of different microorganisms in a single assay. The combination of the obtained data with viability information would represent a significant improvement of the technology. The successful suppression of chaperonin 60 genes (*cpn60*, coding for GroEL) hybridization signals from dead bacteria in defined five-species mixtures as presented in this study appears a promising step in this direction.

*Chapter 5*

# EMERGENCE OF BACTERIAL PATHOGENS

The re-emergence of 'old', that is well known, bacterial pathogens as threat for human health in drinking water can have several reasons that can be attributed either to three complexes: first, the organism; second, the environmental complex including the transmission route; third, the treatment procedure and drinking water supplies; and finally the combined effect of the three complexes (Nwachcuku and Gerba, 2004). For example, on the environmental side, new transmission routes can be opened that can lead to contamination by hitherto not related pathogenic bacteria of the drinking water. Water sources could suffer from higher contamination by treatment-resistant pathogenic bacteria. Global warming will have an impact on growth and survival of pathogenic bacteria including increased growth of pathogens not relevant so far. Additionally, changes in hydrology because of climate change can have severe impacts on source water quantity and quality. Changes in treatment procedures can have a reduced elimination effect on 'old' pathogenic bacteria. Increased loads of organic carbon could promote regrowth of pathogenic bacteria in biofilms and/or the bulk water of the distribution system above beforehand uncrucial level. Despite the multibarrier concept for the drinking water supplies is trying to exclude most risk factors, some factors are difficult to assess. The least predictable risk factor is the drinking water microflora itself. 'Old' pathogens may change, for example they may become more resistant to disinfection procedures, or, the virulence increases and the so far tolerable concentration of the microorganism gives rise to severe health problems. In addition to the well-known pathogens, there are groups of bacteria that are now regarded as 'new' or emerging risk groups for drinking water (Table 1). The most famous ones are Epsilonproteobacteria that have an unrivalled 'career' as waterborne pathogens during the past decade (Miller *et al.*, 2007; Nakagawa *et al.*, 2007). Though not apparent as waterborne pathogen a decade ago, now *Helicobacter pylori*, *Campylobacter jejuni*, and *Campylobacter coli*, and several species of *Arcobacter* are considered waterborne pathogens not only in private wells but also in bulk water and biofilms of public drinking water supplies (Vandenberg *et al.*, 2004; Moreno *et al.*, 2004). Their susceptibility to oxidative disinfection is still under debate and if the Epsilonproteobacteria, as detected by molecular means, are live and infective or dead is not clear (Watson *et al.*, 2004; Vandenberg *et al.*, 2004; Moreno *et al.*, 2004). A proven health risk was shown for private wells in rural areas of developed countries (Sandberg *et al.*, 2006).

The challenge for the safety of the drinking water is to detect pathogenic bacteria that are so far not considered to be of relevance in drinking water, and thus no targeted detection method is possible. One approach to tackle this question is 16S rRNA gene-based fingerprints that provide an overview on the whole bacterial community (Eichler *et al.*, 2006). New members of abundances higher than 0.1% of the total community can be efficiently followed by these fingerprints. Another great advantage of the fingerprint technology is that it can provide an overview on a phylogenetically coherent group of potential pathogens, for example Helicobacteriacea. If group-specific primers are used, the fingerprint shows an overview of different species of the targeted group present in water. The sensitivity and the phylogenetic resolution can be adjusted by an increase in the specificity of the primers used for PCR, for example from the family level to the genus level. Additionally, the phylogenetic resolution can be increased by constructing primers based on the respective fingerprint band sequence and using these for obtaining full 16S rRNA sequence information from environmental DNA (Hofle *et al.*, 2005).

*Chapter 6*

# MOLECULAR VERSUS CULTURABLE METHODS FOR DETECTION

The current standard culture methods used to detect bacteria in the aquatic environment are based on colony-forming units (CFU) counts, thus allowing detection only of bacteria capable of dividing but are unable to detect non growing bacteria and, thus, might not be sufficient for precise monitoring of the microbiological quality of water. However, over the past few years it has been established that nonculturable bacteria, including human bacterial pathogens, represent part of the microbial population in the aquatic environment. These bacteria are usually unable to divide in oligotrophic environments subject to extreme conditions (Roszak and Colwell 1987; Barcina *et al.* 1997; Colwell, 2000). A certain percentage of this population consists of viable bacteria such as the injured forms and the viable but nonculturable (VBNC) forms that are not recoverable in standard culture media (Oliver 1995; Colwell and Huq, 1994). These viable but nonculturable microorganisms exhibit metabolic activity (Rahman *et al.* 1994; Lleo *et al.*, 1998), maintain their pathogenic features (Colwell and Huq, 1994; Pruzzo *et al.*, 2002), maintain their antibiotic resistance traits (Lleo *et al.*, 2003), can express (Lleo *et al.*, 2000) and exchange genes (Arana *et al.*, 1997), and are able to resume division when favorable environmental conditions are replaced (Oliver and Bockian, 1995; Lleo *et al.*, 2001). The VBNC state may be a survival strategy to persist in waters. Because the VBNC cells could make up a potential hazard for human health, it becomes obligatory, for a proper monitoring of the microbiological quality of waters, to develop and apply methods which are also capable of detecting nonculturable bacterial forms. Among the different molecular methods, PCR has proved very useful for detecting low amounts of a specific DNA against a large background of procaryotic and eukaryotic cells and organic material present in environmental samples (Brauns et al., 1991; Leser et al., 1995; Lleo et al., 1999). Moreover, this technique has been modified, thus allowing DNA quantification (cPCR), evaluation of cell viability (mRNA detection by RT-PCR) or visualization of specific cells in a mixed bacterial population (in situ PCR). For these reasons, PCR-based methods are a valid alternative to standard culture methods in that they permit detection of waterborne bacteria in any physiological phase (Brauns et al., 1991; Lleo et al., 1999 and 2000). The use of a molecular method such as PCR could be a valid alternative to detect bacterial faecal contamination indicators such as *Escherichia coli* and *Enterococcus faecalis* and reveal the presence of culturable and nonculturable bacterial forms (Table 2) (Lleo et al., 2005).

The use of a molecular method such as PCR which reveals the presence of specific bacterial DNA, avoids the problem of the detection of nongrowing bacteria with some authors reporting low amplification efficiencies of DNA from VBNC organisms (Brauns et al., 1991) and others amplification efficiency similar in culturable and nonculturable bacteria (Lleo et al., 1999). The molecular methods are the only technique capable of detecting the bacterial population present in waters including the nonculturable subpopulations (injured, dormant, VBNC) which are not detected by culture methods (Lleo et al., 2005). PCR is a method not capable of distinguishing viable and nonviable cells and do not allow the establishment of a correlation between amplified bacterial DNA and viable *Escherichia coli* and *Enterococcus Faecalis* cells. However, because the half-life of the DNA released in the environment is considered to be very short owing to the presence of numerous nucleases in such a complex background (Lorenz and Wackernagel, 1994), it might be deduced that the DNA detected is that contained in nonculturable cells. Moreover, it has been demonstrated (Lleo et al., 2000) by a modified PCR technique, RT-PCR which use the viability indicator mRNA as a target, that in adverse environmental conditions enterococcal cells lose their culturability but are capable of maintaining for several months their viability. The fact that *Escherichia coli* and *Enterococcus faecalis* DNA was detected in only 13% or in 50% of the samples respectively, confirm that is not the free DNA in water which was amplified but that contained in cells which, as known, can persist as intact entities for different periods of time in dependence of their Gram-positive or Gram-negative envelope structure. The nonculturable *Enterococcus faecalis* forms, which constitute part of the aquatic bacterial population and may survive for a long time in waters, can be still viable and, as demonstrated by Pruzzo *et al.* (2002), conserve their ability to adhere to human cells. It was suggested that when the nonculturable bacterial forms present in the environment reach humans via contaminated water and infect the intestinal tract, they are still capable of expressing their adhesive properties (Lleo *et al.*, 2005). Some of these adhered VBNC cells may recover in these newly found optimal environmental conditions (Lleo *et al.* 2001) and initiate the colonization process. This hypothesis is supported by the observation that in at least 50% of the cases of water-borne diseases identified using standard culture methods there is not detectable causative microbial agent. Although it is suspected that viral agents and unknown agents for which there would still be no detection methods can be involved, It was suggest that at least some of these undiagnosed infective microbial agents could be in the VBNC form and may be undetectable with the culture methods currently in use.

The use of culture-based techniques alone is inadequate for detecting the entire, effective bacterial population present in water and that such techniques are not enough to guarantee satisfactory protection of human health (Lleo *et al.*, 2005). To protect human health it is necessary to develop and use methods which detect the nonculturable as well as culturable bacteria present in water.

*Chapter 7*

# CONCLUSIONS

Worldwide, the threat of waterborne disease remains a problem for water treatment and health authorities. Microbial indicators, in particular the faecal coli-forms, are currently used to determine the relative risk of the presence of microbial pathogens. The direct detection of microbial pathogens in water and wastewaters using current conventional methods is time consuming, work intensive and expensive. Moreover, the use of indicator microorganisms as a replacement for the direct detection of microbial pathogens has many disadvantages. Generally, molecular detection methods are essential for the safety of drinking water technologies. These detection technologies have to be specific, reliable, and sensitive. Molecular methods have been developed to increase the speed of analysis. They are able to achieve a high degree of sensitivity and specificity without the need for a complex cultivation and additional confirmation steps. Consequently, some of these methods permit the detection of specific culturable and/or non-culturable bacteria within hours, instead of the days required with the traditional methods. Several molecular methods applied to the specific detection of waterborne pathogens in waters and drinking waters.

A major limitation is still the detection level. Low detection levels can be reached for most pathogenic bacteria. The problem of regrowth of bacteria asks for orders of magnitude lower detection levels unless the tap is included in the sampling scheme. Another major problem for the molecular detection is differentiation between live and dead bacteria. More specifically, it has to be ruled out that pathogenic bacteria can recover under all possible and appropriate circumstances. Though many procedures for assessment of the physiological state of bacteria are available, the development of a reliable procedure for bacterial viability assessment needs further research.

PCR, among the newer methods being investigated, holds great potential for the direct detection of microbial pathogens in water and wastewater. Specific nucleic acid primers already exist for most of the major waterborne pathogens and have been proven to be specific for these organisms. PCR is both highly specific and sensitive and is capable of detecting very small numbers of microorganisms in a sample. In addition, multiple primers can be used to detect different pathogens in one multiplex reaction. PCR does not require the culturing of microorganisms and therefore can improve detection efficiency, time and work. It excludes the requirement for indicator organisms as pathogenic microorganisms can be directly detected from a water or wastewater sample. PCR also has the advantage of being able to be used to determine the viability of a microorganism and thus, is not restricted by dormancy status or the ability to culture the microorganism.

The polymerase enzyme is sensitive to a number of environmental contaminants which can be commonly found in water and wastewaters. Some of these environmental contaminants, such as humic compounds, can also reduce the extraction efficiency of nucleic acids from water samples. The great sensitivity of PCR also makes it susceptible to false positive results due to contamination from extraneous naked nucleic acid, either from the water sample or from the laboratory.

The other major problem associated with the direct detection of microbial pathogens in water and wastewater, which can have an effect on the efficiency of PCR, is the recovery of microbial pathogens from water samples. The use of PCR as a routine surveillance tool in the water industry still remains a potential for the future. At present, the costs and expertise required to use these techniques remain prohibitive for most laboratories. A greater level of automation and improved methods in the recovery of microorganisms, nucleic acid extraction and quantitation would be required before PCR would be accepted in an industry striving to maintain a high degree of water quality for the general public. Rapid advances, however, have recently been made in a number of these problem areas, promising the potential for viable solutions in the near future. Also, as methods such as PCR become more accepted in the fields of microbiology, public health and epidemiology, staff in testing laboratories will become more familiar and proficient in these methods allowing them to be used on a more routine basis.

In the future, greater demands for the provision of safe drinking water, as well as the desire for low levels of risk in wastewater reuse, means improved detection of microbial pathogens in water and wastewater will be essential. PCR has the potential to be one of the quickest and most sensitive of the methods available for microbial pathogen detection. Many detection technologies were developed recently, validation of specific detection methodologies is needed and their application has to be established in laboratories as well as in legal frameworks.

# REFERENCES

Aarnisalo, L. K., Autio, T. J., Lunden, J. M., Saarela, M. H., Korkeala, H. J. & Suihko, M. L. (1999). Subtyping of *Listeria monocytogenes* isolates from food industry with an automated riboprinter microbial characterization system and pulse field gel electrophoresis (PFGE), VTT Biotechnology. VTT Technical Research Centre of Finland.

Abo-Amer, A. E., Soltan, E. M. & Abu-Gharbia M. A. (2008). Molecular approach and bacterial quality of drinking water of urban and rural communities in Egypt. *Acta Microbiol et Immunol. Hungarica*, *55*, 311-326.

Abramowitz, S. (1996) Towards inexpensive DNA diagnostics. *Trends Biotechnol.*, *14*, 397-401

Albert, M. J., Islam, D., Nahar, S., Qadri, F., Falklind, S. & Weintraub, A., (1997). Rapid detection of *Vibrio cholerae* O139 Bengal from stool specimens by PCR. *J. Clin. Microbiol. 35*, 1633-1635.

Alifano, P., Bruni, C. B. & Carlomagno, M. S. (1994). Control of mRNA and decay in prokaryotes. *Genetica* , *94*, 157-172.

Alvarez, J., Porwollik, S., Laconcha, I., Gisakis, V., Vivanco, A. B., Gonzalez, I., Echenagusia, S., Zabala, N., Blackmer, F., McClelland, M., Rementeria, A. & Javier Garaiza, J. (2003). Detection of a *Salmonella enterica* Serovar California Strain Spreading in Spanish Feed Mills and Genetic Characterization with DNA Microarrays. *Appl. Environ. Microbiol.*, *69*, 7531-7534.

Amann, R., Glockner, F. K. & Neef, A. (1997). Modern methods in subsurface microbiology: in situ identification of microorganisms with nucleic acid probes. *FEMS Microbiol. Rev.*, *20*, 191-*200*.

Amann, R. I., Ludwig, W. & Schleifer, K. H. (1995). Phylogenetic identification and in situ detection of individual microbial cells without cultivation. *Microbiol. Rev.*, *59*, 143-169.

American Society for Microbiology (1999). *Microbial pollutants in our nation's water*. American Society for Microbiology, Washington, D.C. 16.

Arana, I., Justo, J. L., Muela, A., Pocino, M., Iriberri, J. & Barcina, I. (1997). Influence of a survival process in a freshwater system upon plasmid transfer between *Escherichia coli* strains. *Microb. Ecol.*, *33*, 41-49.

Aridgides, L. J., Doblin, M. A., Berke, T., Dobbs, F. C., Matson, D. O. & Drake, L. A. (2004). Multiplex PCR allows simultaneous detection of pathogens in ships' ballast water. *Mar. Pollut. Bull.*, *48*, 1096-1101.

Armitage, B. A. (2003). The impact of nucleic acid secondary structure on PNA hybridization. *Drug Discov. Today*, *8*, 222-228.

Azevedo, N. F., Vieira, M. J. & Keevil, C. W. (2003). Establishment of a continuous model system to study *Helicobacter pylori* survival in potable water biofilms. *Water Sci. Technol.*, *47*, 155-160.

Bakau, B. (1993). Regulation of the *Escherichia coli* heat-shock response. *Mol. Microbiol.*, 9, 671-680.

Barcina, I., Lebaron, P. & Vives-Rego, J. (1997). Survival of allochtonous bacteria in aquatic systems: a biological approach. *FEMS Microbiol. Ecol.*, *23*, 1-9.

Barer, M. R., Gribbon, L. T., Harwood, C. R. & Nwoguh, C. E. (1993). The viable but nonculturable hypothesis and medical bacteriology. *Rev. Med. Microbiol.*, *4*, 183-191.

Bartram, J., Chartier, Y., Lee, J. V., Pond, K. & Surman-Lee, S. (2007) *Legionella and the Prevention of Legionellosis*. Geneva, Switzerland: WHO Press, World Health Organization.

Bayardelle, P. & Zafarullah, M. (2002). Development of oligonucleotide primers for the specific PCR-based detection of the most frequent *Enterobacteriaceae* species DNA using *wec* gene templates. *Can. J. Microbiol.*, *48*, 113-122.

Bej, A. K. (2003). Molecular based methods for the detection of microbial pathogens in the environment. *J. Microbiol. Methods*, *53*, 139-140.

Bej, A. K., Di Cesare, J. L., Haff, L. & Atlas, R. M. (1991a). Detection of *Escherichia coli* and *Shigella* spp. in water by using the polymerase chain reaction and gene probes for *uid*. *Appl. Environ. Microbiol.*, *57*, 1013-1017.

Bej, A. K., Mahbubani, M. H. & Atlas, R. M. (1991c). Detection of viable *Legionella pneumophila* in water by polymerase chain reaction and gene probe methods. *Appl. Environ. Microbiol., 57,* 597-600.

Bej, A. K., Mahbubani, M. H., DiCesare, J. L. & Atlas, R. M. (1991d). Polymerase chain reaction-gene probe detection of microorganisms by using filter-concentrated samples. *Appl. Environ. Microbiol., 57,* 3529-3534.

Bej, A. K., McCarty, S. C. & Atlas, R. M. (1991b). Detection of coliform bacteria and *Escherichia coli* by multiple polymerase chain reaction: comparison with defined substrate and plating methods for water quality monitoring. *Appl. Environ. Microbiol., 57,* 2429-2432.

Bej, A. K., Ng, W. Y., Morgan, S., Jones, D. & Mahbubani, M. H. (1996). Detection of viable *Vibrio cholerae* by reverse-transcriptase polymerase chain reaction (RT-PCR). *Mol. Biotechnol., 5,* 1-10.

Bej, A. K., Steffan, R. J., DiCesare, J., Haff, L. & Atlas, R. M. (1990). Detection of coliform bacteria in water by polymerase chain reaction and gene probes. *Appl. Environ. Microbiol., 56,* 307-314.

Bekal, S., Brousseau, R., Masson, L., Prefontaine, G., Fairbrother, J. & Harel, J. (2003). Rapid identification of *Escherichia coli* pathotypes by virulence gene detection with DNA microarrays. *J. Clin. Microbiol., 41,* 2113-2125.

Belasco, J. (1993). mRNA degradation in prokaryotic cells: an overview. In J. Belasco & Brawerman, G. (Eds.), *Control of messenger RNA stability* (pp. 3-12). Inc., San Diego, Calif, Academic Press,

Belgrader, P. Benett, W., Hadley, D., Richards, J., Stratton, P., Mariella Jr., R. and Milanovich, F. (1999). PCR detection of bacteria in seven minutes. *Science, 284,* 449-450.

Bellin, T., Pulz, M., Matussek, A., Hempen, H. G. & Gunzer, F. (2001). Rapid detection of enterohemorrhagic *Escherichia coli* by real-time PCR with fluorescent hybridization probes. *J. Clin. Microbiol., 39,* 370-374.

Berg, G. (1978). The indicator system. In G. Berg (Ed.), *Indictors of Viruses in Water and Food,* 1-13, Ann Arbor Science Publishers, Ann Arbor, MI.

Berney, M., Weilenmann, H. U. & Egli, T. (2006). Flow-cytometric study of vital cellular functions in *Escherichia coli* during solar disinfection (SODIS). *Microbiol., 152,* 1719-1729.

Bernhard, A. E. & Field, K. G. (2000b). A PCR assay to discriminate human and ruminant feces on the basis of host differences in Bacteroides Prevotella genes encoding 16S rRNA. *Appl. Environ. Microbiol., 66,* 4571-4574.

Bernhard, A. E. & Field, K.G. (2000a). Identification of nonpoint sources of fecal pollution in coastal waters by using hostspecific 16S ribosomal DNA genetic markers from fecal anaerobes. *Appl. Environ. Microbiol.*, *66*, 1587-1594.

Blackwood, C. B., Marsh, T., Kim, S. H. & Paul, E. A. (2003). Terminal restriction fragment length polymorphism data analysis for quantitative comparison of microbial communities. *Appl. Environ. Microbiol.*, *69*, 926-932.

Blanch, A. R., Belanche-Munoz, L., Bonjoch, X., Ebdon, J., Gantzer, C., Lucena, F., Ottoson, J., Kourtis, C., Iversen, A. & Kuhn, I. (2006). Integrated analysis of established and novel microbial and chemical methods for microbial source tracking. *Appl Environ Microbiol.*, *72*, 5915-5926.

Bosch, L., Kraal, B., Van der Meide, P. H., Duisterwinkel, F. J. &. Van Noort, J. M. (1983). The elongation factor EF-Tu and its encoding genes. *Prog. Nucleic Acid Res. Mol. Biol.*, *30*, 91-126.

Braganca, S. M., Azevedo, N. F., Simoes, L. C., Keevil, C. W. & Vieira, M. J. (2007). Use of fluorescent in situ hybridization for the visualization of *Helicobacter pylori* in real drinking water biofilms. *Water Sci Technol.*, *55*, 387-393.

Brauns, L. A., Hudson, M. C. & Oliver, J. D. (1991). Use of PCR in detection of culturable and nonculturable *Vibrio vulnificus* cells. *Appl Environ Microbiol.*, *57*, 2651-2655.

Brettar, I. & Hofle, M. G. (1992). Influence of ecosystematic factors on survival of Escherichia coli after large-scale release into lake water mesocosms. *Appl. Environ. Microbiol.*, *58*, 2201-2210.

Brightwell, G., Mowat, E., Clemens, R., Boerema, J., Pulford, D. J. & On, S. L. (2007) Development of a multiplex and real time PCR assay for the specific detection of *Arcobacter butzleri* and *Arcobacter cryaerophilus*. *J. Microbiol. Methods*, *68*, 318-325.

Broczyk, A., Thompson, S., Smith, D. & Lior, H. (1987). Water-borne outbreak of *Campylobacter laridis*-associated gastroenteritis. *Lancet*, *1*, 164-165.

Brousseau, R., Hill, J. E., Prefontaine, G., Goh, S. H., Harel, J. & Hemmingsen, S. M. (2001). *Streptococcus suis* serotypes characterized by analysis of chaperonin 60 gene sequences. *Appl. Environ. Microbiol.*, *67*, 4828-4833.

Burtscher, C., Fall, P.A., Wilderer, P. A. & Wuertz, S. (1999). Detection of *Salmonella* spp. and *Listeria monocytogenes* in suspended organic waste by nucleic acid extraction and PCR. *Appl. Environ. Microbiol.*, *65*, 2235-2237.

Buswell, C. M., Herlihy, Y. M., Lawrence, L. M., McGuiggan, J. T. M., Marsh, P. D., Keevil, C. W. & Leach, S. A. (1998). Extended survival and persistence of Campylobacter spp. in water and aquatic biofilms and their detection by immunofluorescent-antibody and -rRNA staining. *Appl. Environ. Microbiol.*, *64*, 733-741.

Call, D. R., Borucki, M. K. & Loge, F. J. (2003). Detection of bacterial pathogens in environmental samples using DNA microarrays. *J. Microbiol. Methods*, *53*, 235-243.

Call, D. R., Brockman, F. J. & Chandler, D. P. (2001). Detecting and genotyping *Escherichia coli* O157: H7 using multiplexed PCR and nucleic acid microarrays. *Int. J. Food Microbiol.*, *67*, 71-80.

Carson, C. A., Shear, B. L., Ellersieck, M. R. & Asfaw, A. (2001). Identification of fecal *Escherichia coli* from humans and animals by ribotyping. *Appl. Environ. Microbiol.*, *67*, 1503-1507.

Carson, C. A., Shear, B. L., Ellersieck, M. R. & Schnell, J. D. (2003). Comparison of ribotyping and repetitive extragenic palindromic-PCR for identification of fecal *Escherichia coli* from humans and animals. *Appl. Environ. Microbiol.*, *69*, 1836-1839.

Cawthorn, D. M. & Witthuhn, R. C. (2008). Selective PCR detection of viable *Enterobacter sakazakii* cells utilizing propidium monoazide or ethidium bromide monoazide. *J. Appl. Microbiol.*, *105*, 1178-1185.

Chilton, P., Isaacs, N. S., Manas, P. & Mackey, B. M. (2001) Biosynthetic requirements for the repair of membrane damage in pressuretreated *Escherichia coli. Int. J. Food Microbiol.*, *71*, 101-104.

Chizhikov, V., Rasooly, A., Chumakov, K. & Levy, D. D. (2001). Microarray analysis of microbial virulence factors. *Appl. Environ. Microbiol.*, *67*, 3258-3263.

Chizhikov, V., Wagner, M., Ivshina, A., Hoshino, Y., Kapikian, A. Z. & Chumakov, K. (2002). Detection and genotyping of human group A rotaviruses by oligonucleotide microarray hybridization. *J. Clin. Microbiol.*, *40*, 2398-2407.

Cho, J. C. & Tiedji, J. M. (2002). Quantitative detection of microbial genes by using DNA microarrays. *Appl. Environ. Microbiol.*, *68*, 1425-1430.

Chow, K. H., Ng, T. K., Yuen, K. Y. & Yam, W. C. (2001). Detection of RTX toxin gene in *Vibrio cholerae* by PCR. *J. Clin. Microbiol.*, *39*, 2594-2597.

Chowdhury, M. A., Hill, R. T. & Colwell, R. R. (1994). A gene for the enterotoxin zonula occludens toxin is present in *Vibrio mimicus* and *Vibrio cholerae* O139. *FEMS Microbiol. Lett.*, *119*, 377-380.

Colwell, R. R. & Huq, A. (1994). Vibrios in the environment: viable but nonculturable *Vibrio cholerae.* In T. Kaye (Ed.), *Vibrio cholerae and Cholera: Molecular Global Perspectives* (117-133). Washington, DC: American Society for Microbiology.

Colwell, R. R., Brayton, P., Herrington, D., Tall, B., Huq, A. & Levine, M. M. (1996). Viable but non-culturable *Vibrio cholerae* O1 revert to a cultivable state in the human intestine. *World J. Microbiol. Biotechnol.*, *12*, 28-31.

Cook, K. L., Layton, A. C., Dionisi, H. M., Fleming, J. T. & Sayler, G. S. (2004). Evaluation of a plasmid-based 16S–23S rDNA intergenic spacer region array for analysis of microbial diversity in industrial wastewater. *J. Microbiol. Methods*, *57*, 79-93.

Davis, B. D., Luger, S. M. & Tai., P. C. (1986). Role of ribosome degradation in the death of starved *Escherichia coli* cells. *J. Bacteriol.*, *166*, 439-445.

De, K., Ramamurthy, T., Ghose, A. C., Islam, M. S., Takeda, Y., Nair, G. B. & Nandy, R. K. (2001). Modification of the multiplex PCR for unambiguous differentiation of the El Tor and classical biotypes of *Vibrio cholerae* O1. *Indian J. Med. Res.*, *114*, 77-82.

Demidov, V. V., Potaman, V. N., Frank-Kamenetskii, M. D., Egholm, M., Buchard, O., Sonnichsen, S. H. & Nielsen, P. E., (1994). Stability of peptide nucleic acids in human serum and cellular extracts. *Biochem. Pharmacol.*, *48*, 1310-1313.

Di Pinto, A., Ciccarese, G., Tantillo, G., Catalano, D. & Forte, V. T. (2005). A collagenase-targeted multiplex PCR assay for identification of *Vibrio alginolyticus*, *Vibrio cholerae*, and *Vibrio parahaemolyticus. J. Food Prot.*, *68*, 150-153.

Dombek, P. E., Jonhson, L. K., Zimmerley, S. T. & Sadowsky, M. J. (2000). Use of repetitive DNA sequences and the PCR to differentiate Escherichia coli isolates from human and animal sources. *Appl. Environ. Microbiol.*, *66*, 2572-2577.

Egholm, M., Buchardt, O., Christensen, L., Behrens, C., Freier, S. M., Driver, D. A., Berg, R. H., Kim, S. K., Norden, B. & Nielsen, P. E. (1993). PNA hybridizes to complementary oligonucleotides obeying the Watson–Crick hydrogen-bonding rules. *Nature*, *365*, 566-568.

Eichler, S., Christen, R., Holtje, C., Westphal, P., Botel, J., Brettar, I., Mehling, A. & Hofle, M. G. (2006) Composition and dynamics of bacterial communities of a drinking water supply system as assessed by RNA- and DNA-based 16S rRNA gene fingerprinting. *Appl. Environ. Microbiol.*, *72*, 1858-1872.

Erickson, J. W., Vaughn, V., Walter, W. A., Neidhart, F. C. & Gross, C. A.. (1987). Regulation of the promoters and transcripts of *rpoH*, the *Escherichia coli* heat-shock regulatory gene. *Genes Dev.*, *1*, 419-432.

Farag, A. M., Goldstein, J. N., Woodward, D. F. & Samadpour, M. (2001). Water quality in three creeks in the backcountry of Grand Teton National Park, USA. *Journal of Freshwater Ecology*, *16*, 135-143.

Farber, J. M. (1996). An Introduction to the hows and whys of molecular typing. *Journal of Food Protection*, *59*, 1091-1101.

Fey, A., Eichler, S., Flavier, S., Christen, R., Hofle, M. G. & Guzman, C. A. (2004): Development of a real-time PCR-based approach for accurate quantification of bacterial RNA targets in water, using Salmonella as a model organism. *Appl. Environ. Microbiol.*, *70*, 3618-3623.

Fields, P. I., Popovic, T., Wachsmuth, K. & Olsvik, O. (1992). Use of polymerase chain reaction for detection of toxigenic *Vibrio cholerae* O1 strains from the Latin American cholera epidemic. *J. Clin. Microbiol.*, *30*, 2118-2121.

Flekna, G., Stefanic, P., Wagner, M., Smulders, F. J., Mozina, S. S. & Hein, I. (2007). Insufficient differentiation of live and dead *Campylobacter jejuni* and Listeria monocytogenes cells by ethidium monoazide (EMA) compromises EMA/real-time PCR. *Res. Microbiol.*, *158*, 405-412.

Fong, T. & Lipp, E. K. (2005). Enteric viruses of humans and animals in aquatic environments: health risks, detection, and potential water quality assessment tools. *Microbiol. And Molec. Biol. Rev.*, *69*, 357-371.

Fout, G. S., Martinson, B. C., Moyer, M. W. N. & Dahling, D. R. (2003). A multiplex reverse transcription-PCR method for detection of human enteric viruses in groundwater. *Appl. Environ. Microbiol.*, *69*, 3158-3164.

Foy, C. A. & Parkes, H. C. (2001). Emerging homogeneous DNA-based technologies in the clinical laboratory. *Clin. Chem.*, *47*, 990-1000.

Frahm, E. & Obst, U. (2003). Application of the fluorogenic probe technique (TaqMan PCR) to the detection of *Enterococcus* spp. and *Escherichia coli* in water samples. *J. Microbiol. Methods*, *52*, 123-131.

Franck, S. M., Bosworth, B. T. & Moon, H. W. (1998). Multiplex PCR for enteropathogenic, attaching and effacing, and shiga-toxinproducing *E. coli* strains from calves. *J. Clin. Microbiol.*, *36*, 1795-1797.

Fuchs, B. M., Wallner, G., Beisker, W., Schwippl, I., Ludwig, W. & Amann, R., (1998). Flow cytometric analysis of the in situ accessibility of

*Escherichia coli* 16S rRNA for fluorescently labeled oligonucleotide probes. *Appl. Environ. Microbiol., 64*, 4973-4982.

Fukushima, H., Tsunomori, Y. & Seki, R. (2003). Duplex real-time SYBR green PCR assays for detection of 17 species of foodor waterborne pathogens in stools. *J. Clin. Microbiol., 41*, 5134-5146.

Fukushima, M., Kakinuma, K. & Kawaguchi, R. (2002). Phylogenetic analysis of *Salmonella, Shigella*, and *E. coli* strains on the basis of the *gyrB* gene sequence. *J. Clin. Microbiol., 40*, 2779-2785.

Gajardo, R., Pinto, R. M. & Bosch, A. (1995). Polymerase chain reaction amplification and typing of Rotavirus in environmental samples. *Water Sci. Technol., 31*, 371-374.

Giangaspero, A., Cirillo, R., Lacasella, V., Lonigro, A., Marangi, M., Cavallo, P., Berrilli, F., Cave, D. D. & Brandonisio, O. (2009). *Giardia* and *Cryptosporidium* in inflowing water and harvested shellfish in a Lagoon in Southern Italy. *Parasitol. International., 58*, 12-17.

Gibson, U. E. M., Heid, C. A. & Williams, P. M. (1996) A novel method for real-time quantitative RT-PCR. *Genome Res., 6*, 995-1001.

Goh, S. H., Potter, S., Wood, J. O., Hemmingsen, S. M., Reynolds, R. P. & Chow, A. W. (1996). HSP60 gene sequences as universal targets for microbial species identification: studies with coagulase-negative staphylococci. *J. Clin. Microbiol., 34*, 818-823.

Grabow, W. O. K. (1996). Waterborne diseases: Update on water quality assessment and control. *Water SA., 22*, 193-202.

Guan, S., Xu, R., Chen, S., Odumeru, J. & Gyles, C. (2002). Development of a procedure for discriminating among *E. coli* isolates form animal and human sources. *Appl. Environ. Microbiol., 68*, 2690-2698.

Gubala, A. A. (2006). Multiplex real-time PCR detection of *Vibrio cholerae*. *J. Microbiol. Methods, 65*, 278-293.

Hanninen, M. L., Haajanen, H., Pummi, T., Wermundsen, K., Katila, M. L., Sarkkinen, H., Miettinen, I. & Rautelin, H. (2003). Detection and typing of *Campylobacter jejuni* and *Campylobacter coli* and analysis of indicator organisms in three waterborne outbreaks in Finland. *Appl. Environ. Microbiol., 69*, 1391-1396.

Haramoto, E., Katayama, H. & Ohgaki, S. (2004). Detection of noroviruses in tap water in Japan by means of a new method for concentrating enteric viruses in large volumes of freshwater. *Appl. Environ. Microbiol., 70*, 2154-2160.

Harms, G., Layton, A. C., Dionisi, H. M., Gregory, I. R., Garrett, V. M., Hawkins, S. A., Robinson, K. G. & Sayler, G. S. (2003). Real-time quantification of nitrifying bacteria in a municipal wastewater treatment plant. *Environ. Sci. Technol.*, *37*, 343-351.

Harmsen, H. J., Raangs, G. C., He, T., Degener, J. E. & Welling, G. W. (2002). Extensive set of 16S rRNA-based probes for detection of bacteria in human feces. *Appl. Environ. Microbiol.*, *68*, 2982-2990.

Hay J., Seal D. V., Billcliffe, B. & Freer, J. H. (1995). Non-culturable *Legionella pneumophila* associated with *Acanthamoeba castellanii*: detection of the bacterium using DNA amplification and hybridization. *J. Appl. Bacteriol.*, *78*, 61-65.

Hayashi, H., Sakamoto, M. & Benno, Y. (2002). Phylogenetic analysis of the human gut microbiota using 16S rDNA clone libraries and strictly anaerobic culture-based methods. *Microbiol. Immunol.*, *46*, 535-548.

Head, I. M., Saunders, J. R. & Pickup, R. W. (1998). Microbial evolution, diversity, and ecology: a decade of ribosomal RNA analysis of uncultivated microorganisms. *Microb. Ecol.*, *35*, 1-21.

Heid, C. A., Stevens, J., Livak, K. J. & Williams, P. M. (1996). Real-time quantitative PCR. *Genome Res.*, *6*, 986-994.

Heidelberg, J. F., Eisen, J. A., Nelson, W. C., Clayton, R. A., Gwinn, M. L., Dodson, R. J., Haft, D. H., Hickey, E. K., Peterson, J. D., Umayam, L., Gill, S. R., Nelson, K. E., Read, T. D., Tettelin, H., Richardson, D., Ermolaeva, M. D., Vamathevan, J., Bass, S., Qin, H., Dragoi, I., Sellers, P., McDonald, L., Utterback, T., Fleishmann, R. D., Nierman, W. C. & White, O. (2000). DNA sequence of both chromosomes of the cholera pathogen *Vibrio cholerae*. *Nature*, *406*, 477-483.

Hein, I., Lehner, A., Rieck, P., Klein, K., Brandl, E. & Wagner, M. (2001). Comparison of different approaches to quantify *Staphylococcus aureus* cells by real-time quantitative PCR and application of this technique for examination of cheese. *Appl. Environ. Microbiol.*, *67*, 3122-3126.

Hendrick, J. P. & Hartl, F. U. (1993). Molecular chaperone functions of heat-shock proteins. *Annu. Rev. Biochem.*, *62*, 349-384.

Hill, J. E., Seipp, R. P., Betts, M., Hawkins, L. L., Van Kessel, A. G. A. G., Crosby, W. L. & Hemmingsen, S. M. (2002). Extensive profiling of a complex microbial community by high-throughput sequencing. *Appl. Environ. Microbiol.*, *68*, 3055-3066.

Hill, V. R., Kahler, A. M., Jothikumar, N., Johnson, T. B., Hahn, D. & Cromeans, T. L. (2007) Multistate evaluation of an ultrafiltration-based procedure for simultaneous recovery of enteric microbes in 100-liter tap water samples. *Appl. Environ. Microbiol.*, *73*, 4218-4225.

Hodson, R. E., Dustman, W. A., Garg, R. P. & Moran, M. A. (1995). In situ PCR for visualization of microscale distribution of specific genes and gene products in prokaryotic communities. *Appl. Environ. Microbiol.*, *6*, 4074-4082.

Hoefel, D., Monis, P. T., Grooby, W. L., Andrews, S. & Saint, C. P. (2005) Culture independent techniques for rapid detection of bacteria associated with loss of chloramine residual in a drinking water system. *Appl Environ Microbiol.*, *71*, 6479-6488.

Hofle, M. G., Flavier, S., Christen, R., Botel, J., Labrenz, M. & Brettar, I. (2005). Retrieval of nearly complete 16S rRNA gene sequences from environmental DNA following 16S rRNA-based community fingerprinting. *Environ. Microbiol.*, *7*, 670-675.

Holland, P. M., Abramson, R. D., Watson, R. & Gelfand, D. H. (1991). Detection of specific polymerase chain reaction product by utilizing the 5'-3' exonuclease activity of *Thermus aquaticus* DNA polymerase. *Proc. Natl. Acad. Sci.*, USA, 88, 7276-7280.

Holloway, P. (2001). Tracing the source of *E. coli* fecal contamination of water using rep-PCR, Manitoba Livestock Manure Management Initiative Project: MLMMI 00-02-08 . University of Winnipeg.

Hong. J., Jung, W. K., Kim, J. M., Kim, S. H., Koo, H. C., Ser, J. & Park, Y. H. (2007). Quantification and differentiation of Campylobacter jejuni and Campylobacter coli in raw chicken meats using a real-time PCR method. *J. Food. Prot.*, *70*, 2015-2022.

Hoorfar, J., Ahrens, P. & Radstrom, P. (200). Automated 50 Nuclease PCR Assay for Identification of Salmonella enterica. *J. Clin. Microbiol.*, *38*, 3429-35.

Horman, A., Rimhanen-Finne, R., Maunula, L., von Bonsdorff, C. H., Torvela, N., Heikinheimo, A. & Hanninen, M. L. (2004). *Campylobacter* spp., *Giardia* spp., *Cryptosporidium* spp., Noroviruses, and indicator organisms in surface water in southwestern Finland, 2000-2001. *Appl. Environ. Microbiol.*, *70*, 87-95.

Hoshino, K., Yamasaki, S., Mukhopadhyay, A. K., Chakraborty, S., Basu, A., Bhattacharya, S. K., Nair, G. B., Shimada, T. & Takeda, Y. (1998). Development and evaluation of a multiplex PCR assay for rapid detection of toxigenic *Vibrio cholerae* O1 and O139. *FEMS Immunol. Med. Microbiol.*, *20*, 201-207.

Hsu, B., Ma, P., Tai-Sheng Liou, T., Chen, J. & Shih, F. (2009). Identification of 18S ribosomal DNA genotype of *Acanthamoeba* from hot spring recreation areas in the central range, Taiwan. *J. Hydrol.*, *1367*, 249-254.

Huijsdens, X. W., Linskens, R. K., Mak, M., Meuwissen, S. G., Vandenbroucke-Grauls, C. M. & Savelkoul, P. H. (2002). Quantification of bacteria adherent to gastrointestinal mucosa by realtime PCR. *J. Clin. Microbiol.*, *40*, 4423-4427.

Huq, A., Rivera, I. N. G. & Colwell, R. R. (2000). Epidemiological significance of viable but non culturable microorganisms. In R. R. Colwell, & D. J. Grimes (Eds.), *Nonculturable Microrganisms in the Environment* (301-323). Washington, DC: ASM Press.

Ibekwe, A. M., Watt, P. M., Grieve, C. M., Sharma, V. K. & Lyons, S. R. (2002). Multiplex fluorogenic real-time PCR for detection and quantification of *Escherichia coli* O157:H7 in dairy wastewater wetlands. *Appl. Environ. Microbiol.*, *68*, 4853-4862.

Iqbal, S., Robinson, J., Deere, D., Saunders, J. R., Edwards, C. & Porter, J. (1997). Efficiency of the polymerase chain reaction amplification of the uid gene for detection of *Escherichia coli* in contaminated water. *Lett. Appl. Microbiol.*, *24*, 498-502.

Jeffrey, W. H., Nazaret, S. & Von Haven, R. (1994). Improved method for recovery of mRNA from aquatic samples and its application to detection of *mer* expression. *Appl. Environ. Microbiol.*, *60*, 1814-1821.

Jothikumar, N. & Griffiths, M. W. (2002). Rapid Detection of *Escherichia coli* O157:H7 with Multiplex Real-Time PCR Assays. *Appl. Environ. Microbiol.*, *68*, 3169-3171.

Jothikumar, N., Wang, X. & Griffiths, M. W. (2003). Real-time multiplex SYBR green I-based PCR assay for simultaneous detection of *Salmonella serovars* and *Listeria monocytogenes*. *J. Food Prot.*, *66*, 2141-2145.

Joux, F. & Lebaron, P. (2000). Use of fluorescent probes to assess physiological functions of bacteria at single-cell level. *Microbes Infect.*, *2*, 1523-1535.

Juck, D., Ingram, J., Prevost, M., Coallier, J. & Greer, C. (1996). Nested PCR protocol for the rapid detection of *Escherichia coli* in potable water. *Can. J. Microbiol.*, *42*, 862-866.

Jury, W. A. & Vaux, H. (2005). The role of science in solving the world's emerging water problems. *Proc. Natl. Acad. Sci.*, *157*, 15715-15720.

Kapley, A & Purohit, H. J. (2001). Detection of etiological agent for cholera by PCR protocol. *Med. Sci. Monit.*, *7*, 242-245

Kapley, A., Lampel, K. & Purohit, H. J. (2000a). Thermocycling steps and optimization of multiplex PCR. *Biotechnol. Lett.*, *22*, 1913-1918.

Kapley, A., Lampel, K. & Purohit, H. J. (2000b). Development of duplex PCR for *Salmonella* and *Vibrio*. *World J. Microbiol. & Biotechnol.*, *16*, 457-58

Kapley, A., Lampel, K. & Purohit, H. J. (2001). Rapid detection of *Salmonella* in water samples by Multiplex PCR. *Water Environ. Res.*, *73*, 461-465.

Kaprelyants, A. S., Gottschal, J. C. & Kell, D. B. (1993). Dormancy in non-sporulating bacteria. *FEMS Microbiol. Rev.*, *104*, 271-286.

Keasler, S. P. & Hall, R. H. (1993). Detecting and biotyping Vibrio cholerae O1 with multiplex polymerase chain reaction. *Lancet*, *341*, 1661.

Keramas, G., Bang, D. D., Lund, M., Madsen, M., Rasmussen, S. E., Bunkenborg, H., Tellemanm, P. & Christensen, C. B. V. (2003). Development of a sensitive DNA microarray suitable for rapid detection of *Campylobacter* spp. *Mol. Cell. Probes*, *17*, 187-196.

Klein, P. G. & Juneja, V. K. (1997). Sensitive detection of viable *Listeria monocytogenes* by reverse-transcriptase PCR. *Appl. Environ. Microbiol.*, *63*, 4441-4448.

Kong, R. Y. C. , Lee, S. K. Y., Law, T. W. F., Law, S. H. W. & Wu, R. S. S. (2002) Rapid detection of six types of bacterial pathogens in marine waters by multiplex PCR. *Water Research*, *36*, 2802-2812.

Kong, R. Y. C., Dung, W. F., Vrijmoed, L. L. P. & Wu, R. S. S. (1995) Codetection of three species of waterborne bacteria by multiplex PCR. *Mar. Pollut. Bull.*, *31*, 17-24.

Kong, R. Y. C., So, C. L., Law, W. F. & Wu, R. S. S. (1999) A sensitive and versatile multiplex PCR system for the rapid detection of enterotoxigenic (ETEC), enterohaemorrhagic (EHEC) and enteropathogenic (EPEC) strains of *Escherichia coli*. *Mar. Pollut. Bull.*, *38*, 1207-15.

Kowalchuk, G. A., de Bruijn, F. J., Head, I. M., Akkermans, A. D. & van Elsas, J. D. (2004). *Molecular Microbial Ecology Manual*. Springer-Verlag.

Kreader, C. A. (1996). Relief of amplification inhibition in PCR with bovine serum albumin or T4 gene 32 protein. *Appl. Environ. Microbiol.*, *62*, 1102-1106.

Kutyavin, I. V., Afonina, I. A., Mills, A., Gorn, V. V., Lukhtanov, E. A., Belousov, E. S., Singer, M. J., Walburger, D. K., Lokhov, S. G., Gall, A. A., Dempcy, R., Reed, M. W., Meyer, R. B. & Hedgpeth, J. (2000). 3V-minor groove binder-DNA probes increase sequence specificity at PCR extension temperatures. *Nucleic Acids Res.*, *28*, 655-661.

Laberge, I., Ibrahim, A., Barta, J. R. & Griffiths, M. W. (1996). Detection of *Cryptosporidium parvum* in raw milk by PCR and oligonucleotide probe hybridization. *Appl. Environ. Microbiol.*, *62*, 3259-3264.

Le Guyader, F., Menard, D., Pommepuy, M. & Kopecka, H. (1995). Use of RT seminested PCR to assess viral contamination in caribbean river (Martinique). *Water Sci. Technol.*, *31*, 391-394.

Leclere, H. D., Mossel, A. A., Edberg, S. C. & Struijk, C. B. (2001). Advances in the bacteriology of the coliform group: their suitability as markers of microbial water safety. *Annu. Rev. Microbiol.*, *55*, 201-234.

Lee, C. Y., Panicker, G. & Bej, A. K. (2003). Detection of pathogenic bacteria in shellfish using multiplex PCR followed by CovaLink NH microwell plate sandwich hybridization. *J. Microbiol. Methods*, *53*, 199-209.

Lehtola, M. J., Loades, C. J. & Keevil, C. W. (2005). Advantages of peptide nucleic acid oligonucleotides for sensitive site directed 16S rRNA fluorescence in situ hybridization (FISH) detection of *Campylobacter jejuni*, *Campylobacter coli* and *Campylobacter lari*. *J. Microbiol. Methods*, *62*, 211-219.

Leser, T. D., Boye, M. & Hendriksen, N. B. (1995). Survival and activity of *Pseudomonas* sp strain B13 (FR1) in a marine microcosm determined by quantitative PCR and an rRNA-targeting probe and its effect on the indigenous bacterioplankton. *Appl. Environ. Microbiol.*, *61*, 1201-1207.

Leung, K. T., Mackereth, R., Tien, Y. C. & Topp, E. (2004). A comparison of AFLP and ERIC-PCR analyses for discriminating Escherichia coli from cattle, pig and human sources. *FEMS Microbial. Ecol.*, *47*, 111-119.

Liang, J. L., Dziuban, E. J., Craun, G. F., Hill, V., Moore, M. R., Gelting, R. J., Calderon, R. L., Beach, M. J. & Roy, S. L. (2006) Surveillance for waterborne disease and outbreaks associated with drinking water and water not intended for drinking-United States, 2003-2004. *MMWR Surveill Summ.*, *55*, 5512-5561.

Lipman, L. J. A., de Nijs, A., Lam, T. J. G. M. & Gaastra, W. (1995). Identification of *Escherichia coli* strains from cows with clinical mastitis by serotyping and DNA polymorphism patterns with REP and ERIC primers. *Vet. Microbiol.*, *43*,13-19.

Lipp, E. K., Rivera, I. N., Gil, A. I., Espeland, E. M., Choopun, N., Louis, V. R., Russek-Cohen, E., Huq, A. & Colwell, R. R. (2003). Direct detection of *Vibrio cholerae* and *ctxA* in Peruvian coastal water and plankton by PCR. *Appl. Environ. Microbiol.*, *69*, 3676-3680.

Livak, K. J., Flood, S. J. A., Marmaro, J., Giusti, W. & Deetz, K. (1995). Oligonucleotides with fluorescent dyes at opposite ends provide a quenched probe system useful for detecting PCR product and nucleic acid hybridization. *PCR Methods Appl.*, *4*, 357-362.

Lleo, M. M., Bonato, B., Benedetti, D. & Canepari, P. (2005). Survival of enterococcal species in aquatic environments. *FEMS Microbiol. Ecol.*, *54*, 189-196.

Lleo, M. M., Bonato, B., Signoretto, C. & Canepari, P. (2003). Vancomycin resistance is maintained in enterococci in the viable but nonculturable state and after division has resumed. *Antimicrob. Agents Chemother.*, *47*, 1154-1156.

Lleo, M. M., Bonato, B., Tafi, M. C., Signoretto, C., Boaretti, M. & Canepari, P. (2001). Resuscitation rate in different enterococcal species in the viable but nonculturable state. *J. Appl. Microbiol.*, *91*, 1095-1102.

Lleo, M. M., Pierobon, S., Tafi, M. C., Signoretto, C. & Canepari, P. (2000). mRNA detection by RT-PCR for monitoring viability over time in an *Enterococcus faecalis* viable but nonculturable population maintained in a laboratory microcosm. *Appl. Environ. Microbiol.*, *66*, 4564-4567.

Lleo, M. M., Tafi, M. C. & Canepari, P. (1998). Nonculturable *Enterococcus faecalis* cells are metabolically active and capable of resuming active growth. *Syst. Appl. Microbiol.*, *21*, 333-339.

Lleo, M. M., Tafi, M. C., Signoretto, C., Dal Cero, C. & Canepari, P. (1999). Competitive polymerase chain reaction for quantification of nonculturable *Enterococcus faecalis* cells in lake water. *FEMS Microbiol. Ecol.*, *30*, 345-353.

Lofstrom, C., Knutsson, R., Axelsson, C. E. & Radstrom, P. (2004). Rapid detection of *Salmonella* spp. in animal feed samples by PCR after culture enrichment. *Appl. Environ. Microbiol.*, *70*, 69-75.

Loge, F. J., Thompson, D. E. & Call, D. R. (2002). PCR detection of specific pathogens in water: a risk-based analysis. Environ. *Sci. Technol.*, *36*, 2754-2759.

Lorenz, M. G. & Wackernagel, W. (1994). Bacterial gene transfer by natural genetic transformation in the environment. *Microbiol. Rev.*, *58*, 563-602.

Lubeck, P. S., Cook, N., Wagner, M., Fach, P. & Hoorfar, J. (2003b). Toward an international standard for PCR-based detection of food-borne thermotolerant Campylobacters: validation in a multicenter collaborative trial. *Appl. Environ. Microbiol.*, *69*, 5670-5672.

Lubeck, P. S., Wolffs, P., On, S. L. W., Ahrens, P., Radstrom, P. & Hoorfar, J. (2003a). Toward an international standards for PCR-based detection of food-borne thermotolerant Campylobacters: assay development and analytical validation. *Appl. Environ. Microbiol.*, *69*, 5664-5669.

Lyon, W. J. (2001). TaqMan PCR for detection of *Vibrio cholerae* O1, O139, non-O1, and non-O139 in pure cultures, raw oysters, and synthetic seawater. *Appl. Environ. Microbiol., 67*, 4685-4693.

Lyons, S. R., Griffen, A. L. & Leys, E. J. (2000). Quantitative real-time PCR for Porphyromonas gingivalis and total bacteria. *J. Clin. Microbiol., 38*, 2362-2365.

Malinen, E., Kassinen, A., Rinttila, T. & Palva, A. (2003). Comparison of real-time PCR with SYBR Green 1 or 50-nuclease assays and dot-blot hybridization with rDNA-targeted oligonucleotide probes in quantification of selected faecal bacteria. *Microbiol., 149*, 269-277.

Malorny, B., Hoorfar, J., Bunge, C. & Helmuth, R. (2003). Multicenter validation of the analytical accuracy of *Salmonella* PCR: towards an international standards. *Appl. Environ. Microbiol., 69*, 290-296.

Marre, R., Kwaik, Y. J., Bartlett, C., Cianciotto, N. P., Fields, B. S., Frosch, M., Hacker, J. & Luck, P. C. (2002). *Legionella*. Washington, DC: ASM Press.

Matsuki, T., Watanabe, K., Fujimoto, J., Kado, Y., Takada, T., Matsumoto, K. & Tanaka, R. (2004). Quantitative PCR with 16S rRNA-gene-targeted species-specific primers for analysis of human intestinal bifidobacteria. *Appl. Environ. Microbiol., 70*, 167-173.

Mayer, C. L. & Palmer, C. J. (1996). Evaluation of PCR, nested PCR and fluorescent antibodies for detection of *Giardia* and *Cryptosporidium* species in wastewater. *Appl. Environ. Microbiol., 62*, 2081-2085.

Maynard, C., Berthiaume, F., Lemarchand, K., Harel, J., Payment, P., Bayardelle, P., Masson, L. & Brousseau1, R. (2005). Waterborne Pathogen Detection by Use of Oligonucleotide-Based Microarrays. Applied and Environmental Microbiology. *Appl. Environ. Microbiol., 71*, 8548-8557.

McCabe, K. M., Zhang, Y. H., Huang, B. L., Wagar, E. A. & McCabe, E. R. (1999). Bacterial species identification after DNA amplification with a universal primer pair. *Mol. Genet. Metab., 66*, 205-211.

Mendes, C. L., Abath, F. G. & Leal, N C. (2008). Development of a multiplex single-tube nested PCR (MSTNPCR) assay for *Vibrio cholerae* O1 detection. *J Microbiol Methods, 72*, 191-196.

Miller, S. M., Tourlousse, D. M., Stedtfeld, R. D., Baushke, S. W., Herzog, A. B., Wick, L. M., Rouillard, J. M., Gulari, E., Tiedje, J. M. & Hashsham, S. A. (2008). An in-situ synthesized virulence and marker gene (VMG) biochip for the detection of bacterial pathogens in water. *Appl. Environ. Microbiol., 74*, 2200-2209.

Miller, W. G., Parker, C. T., Rubenfield, M., Mendz, G.L., Wosten, M. M., Ussery, D. W., Stolz, J. F., Binnewies, T. T., Hallin, P. F. & Wang, G. (2007). The complete genome sequence and analysis of the epsilonproteobacterium *Arcobacter butzleri*. *PLoS ONE*, *2*, e1358.

Mitra, R. K., Nandy, R. K., Ramamurthy, T., Bhattacharya, S. K., Yamasaki, S., Shimada, T., Takeda, Y. & Nair, G. B. (2001). Molecular characterisation of rough variants of *Vibrio cholerae* isolated from hospitalised patients with diarrhoea. *J. Med. Microbiol.*, *50*, 268-276.

Monis, P. T., Giglio, S., Keegan1, A. R. & Thompson, R. C. A. (2005). Emerging technologies for the detection and genetic characterization of protozoan parasites. *TRENDS in Parasitol.*, *21*, 340-346.

Moreno, Y. M., Hernandez, M. A., Ferru J. L., Alonso, S., Botella, B., Montes, R. & Hernandez, J. (2001). Direct detection of thermotolerant campylobacters in chicken products by PCR and in situ hybridization. *Res. Microbiol.*, *152*, 577-582.

Moreno, Y., Alonso, J. L., Botella, S., Ferrus, M. A. & Hernandez, J. (2004) Survival and injury of *Arcobacter* after artificial inoculation into drinking water. *Res. Microbiol.*, *155*, 726-730.

Moreno, Y., Botella, S., Alonso, J. L., Ferrus, M. A., Hernandez, M. & Hernandez, J. (2003). Specific detection of *Arcobacter* and *Campylobacter* strains in water and sewage by PCR and fluorescent in situ hybridization. *Appl. Environ. Microbiol.*, *69*, 1181-1186.

Morin, N. J., Gong, Z. & Li, X. F. (2004). Reverse transcription-multiplex PCR assay for simultaneous detection of *Escherichia coli* O157:H7, *Vibrio cholerae* O1, and *Salmonella typhi*. *Clin. Chem.*, *50*, 2037-2044.

Morio, F., Corvec, S., Caroff, N., Le Gallou, F., Drugeon, H. & Reynaud, A. (2007). Real-time PCR assay for the detection and quantification of *Legionella pneumophila* in environmental water samples: utility for daily practice. *Int. J. Hyg. Environ. Health*, *211*, 403-11

Moter, A. & Gobel, U. B. (2000). Fluorescence in situ hybridization (FISH) for direct visualization of microorganisms. *J. Microbiol. Methods*, *41*, 85-112.

Nagashima, K., Hisada, T., Sato, M. & Mochizuki, J. (2003). Application of new primer-enzyme combinations to terminal restriction fragment length polymorphism profiling of bacterial populations in human feces. *Appl. Environ. Microbiol.*, *69*, 1251-1262.

Nakagawa, S., Takaki, Y., Shimamura, S., Reysenbach, A. L., Takai, K. & Horikoshi, K. (2007) Deep sea vent epsilonproteobacterial genomes provide insights into emergence of pathogens. *Proc. Natl. Acad. Sci.*, *104*, 12146-12150.

Nandi, B., Nandy, R. K., Mukhopadhyay, S., Nair, G. B., Shimada, T. & Ghose, A. C. (2000). Rapid method for species-specific identification of *Vibrio cholerae* using primers targeted to the gene of outer membrane protein OmpW. *J. Clin. Microbiol.*, *38*, 4145-4151.

Nayak, A. K. & Rose, J. B. (2007) Detection of *Helicobacter pylori* in sewage and water using a new quantitative PCR method with SYBR green. *J. Appl. Microbiol.*, *103*, 1931-1941.

Nebe-von-Caron, G., Stephens, P. J., Hewitt, C. J., Powell, J. R. & Badley, R. A. (2000). Analysis of bacterial function by multi-colour fluorescence flow cytometry and single cell sorting. *J. Microbiol . Methods.*, *42*, 97-114.

Nguyen, T. V., Van, P. L., Huy, C. L., Gia, K. N. & Weintraub, A. (2005) Detection and characterization of diarrheagenic *Escherichia coli* from young children in Hanoi, Vietnam. *J. Clin. Microbiol.*, *43*, 755-760.

Nocker, A., Cheung, C. Y. & Camper, A. K. (2006). Comparison of propidium monoazide and ethidium monoazide for differentiation of live vs. dead bacteria by selective removal of DNA from dead cells. *J. Microbiol. Methods*, *67*, 310-320.

Nocker, A., Mazza, A., Masson, L., Camper, A. K. & Brousseau, R. (2009). Selective detection of live bacteria combining propidium monoazide sample treatment with microarray technology. *J. Microbiol. Methods*, *76*, 253-261.

Nocker, A., Sossa, K. & Camper, A. K. (2007a). Molecular monitoring of disinfection efficacy. *J. Microbiol. Methods*, *70*, 252-260.

Nocker, A., Sossa, P., Burr, M. & Camper, A. K. (2007b). Use of propidium monoazide for livedead distinction in microbial ecology. *Appl. Environ. Microbiol.*, *73*, 5111-5117.

Nogva, H. K., Bergh, A., Holck, A. & Rudi, K. (2000a) Application of the 50-nuclease PCR assay in evaluation and development of methods for quantitative detection of *Campylobacter jejuni*. *Appl. Environ. Microbiol.*, *66*, 4029-36.

Nogva, H. K., Dromtorp, S. M., Nissen, H. & Rudi, K. (2003). Ethidium monoazide for DNA-based differentiation of viable and dead bacteria by 5′-nuclease. *PCR BioTechniques*, *810*, 804-813.

Nordstrom, J. L., Vickery, M. L. C., Blackstone, G. M., Murray, S. L. & DePaola, A. (2007). Development of a Multiplex Real-Time PCR Assay with an Internal Amplification Control for the Detection of Total and

Pathogenic *Vibrio parahaemolyticus* Bacteria in Oysters. *Appl. Environ. Microbiol.*, *73*, 5840-5847.

Novga, H. K., Rudi, K., Naterstad, K., Holck, A. & Lillehaug, D. (2000b). Application of 50-nuclease PCR for quantitative detection of *Listeria monocytogenes* in pure cultures, water, skim milk, and unpasteurized whole milk. *Appl. Environ. Microbiol.*, *66*, 4266-4271.

Nwachcuku, N. & Gerba, C. P. (2004). Emerging waterborne pathogens: can we kill them all? *Curr. Opin. Biotechnol.*, *15*, 175-180.

OECD, Organization for Economic Cooperation and Development, WHO (2003). Assessing Microbial Safety of Drinking Water: Improving Approaches and Methods. World Health Organization, Organization for Economic Co operation and Development, London, IWA.

Ogunjimi, A. A. & Choudary, P. V. (1999). Adsorption of endogenous polyphenols relieves the inhibition by fruit juices and fresh produce of immuno-PCR detection of *Escherichia coli* O157: H7. *FEMS Immunol. Med. Microbiol.*, *23*, 213-20.

Okabe, S., Satoh, H. & Watanabe, Y. (1999). In situ analysis of nitrifying biofilms as determined by in situ hybridization and the use of microelectrodes. *Appl. Environ. Microbiol.*, *65*, 3182-3191.

Oliver, J. D. & Bockian, R. (1995). In vivo resuscitation, and virulence towards mice, of viable but nonculturable cells of Vibrio vulnificus. *Appl. Environ. Microbiol.*, *61*, 2620-2623.

Ootsubo, M., Shimizu, T., Tanaka, R., Sawabe, T., Tajima, K. & Ezura, Y. (2002). Oligonucleotide probe for detecting Enterobacteriaceae by in situ hybridization. *J. Appl. Microbiol.*, *93*, 60-68.

Panicker, G., Myers, M. L. & Bej, A. K. (2004a). Rapid detection of *Vibrio vulnificus* in shellfish and Gulf of Mexico water by real-time PCR. *Appl. Environ. Microbiol.*, *70*, 498-507.

Parveen, S., Portier, K. M., Robinson, K., Edmiston, L. & Tamplin, M. L. (1999). Discriminant analysis of ribotype profiles of *Escherichia coli* for differentiating human and nonhuman sources of fecal pollution. *Appl. Environ. Microbiol.*, *65*, 3142-3147.

Patel, B. K. R., Banjerjee, D. K. & Butcher, P. D. (1993). Determination of *Mycobacterium leprae* viability by polymerase chain reaction amplification of 71-kDa heat shock protein mRNA. *J. Infect. Dis.*, *168*, 799-800.

Paul, J. H. & Pichard, S. L. (1995). Extraction of DNA and RNA from aquatic environments. In J. T. Trevors, & J. D. van Elsas (Eds.), *Nucleic Acids in the Environment* (153-177). Springer-Verlag.

Pepper, I. L., Straub, T. M. & Gerba C. P. (1997). Detection of microorganisms in soils and sludges. In. G. A. Toranzos (Ed.), *Environmental applications of nucleic acid amplification techniques* (pp.95-111). Technomic Publishing Company, Lancaster, PA.

Pernthaler, A. & Amann, R. (2004). Simultaneous fluorescent in situ hybridization of mRNA and rRNA in environmental bacteria. *Appl. Environ. Microbiol.*, *70*, 5426-5433.

Perry-O Keefe, H., Rigby, S., Olivieira, K., Sbrensen, D., Stender, H., Coull, J. & Hyldig-Nielsen, J. J. (2001). Identification of indicator microorganisms using a standardized PNA FISH method. *J. Microbiol. Methods*, *47*, 281-292.

Pichard, S. L. & Paul, J. H. (1993). Gene expression per gene dose, a specific measure of gene expression in aquatic micro-organisms. *Appl. Environ. Microbiol.*, *59*, 451-457.

Picone, T., Young, T. & Fricker, E. (1997). Detection of *Legionella* and *Legionella pneumophila* in environmental water samples using the polymerase chain reaction. In Toranzos, G. A. (Ed). *Environmental applications of nucleic acid amplification techniques* (pp. 159-182). Technomic Publishing Company, Lancaster, PA.

Pruzzo, C., Tarsi, R., Lleo, M. M., Signoretto, C., Zampini, M., Colwell, R. R. & Canepari, P. (2002). In vitro adhesion to human cells by viable but nonculturable *Enterococcus faecalis*. *Curr. Microbiol.*, *45*, 105-110.

Quintero-Betancourt, W., Peele, W. R. & Rose, J. B. (2002). *Cryptosporidium parvum* and *Cyclospora cayetanensis*: a review of laboratory methods for detection of these waterborne parasites. *J. Microbiol. Methods*, *49*, 209-224

Rahman, I., Shahamat, M., Kirchman, P. A., Rissek-Cohen, E. & Colwell, R. R. (1994). Methionine uptake and cytopathogenicity of viable but nonculturable *Shigella dysenteriae* type 1. *Appl. Environ. Microbiol.*, *60*, 3573-3578.

Rhee, S. K., Liu, X., Wu, L., Chong, S. C., Wan, X. & Zhou, J. (2004). Detection of genes involved in biodegradation and biotransformation in microbial communities by using 50-mer oligonucleotide microarrays. *Appl. Environ. Microbiol.*, *70*, 4303-4317.

Richardson, H. Y., Nichols, G., Lane, C., Iain, R. Lake, I. R. & Hunter, P. R. (2009). Microbiological Surveillance of Private Water Supplies in England–The impact of environmental and climate factors on water quality. *Water research*, In press.

Ritzler, M. & Altwegg, M. (1996). Sensitivity and specificity of a commercially available enzyme-linked immunoassay for the detection of polymerase chain reaction amplified DNA. *J. Microbiol. Methods*, *27*, 233-238.

Rivera, I. N., Chun, J., Huq, A., Sack, R. B. & Colwell, R. R. (2001). Genotypes associated with virulence in environmental isolates of *Vibrio cholerae. Appl. Environ. Microbiol.*, *67*, 2421-2429.

Rivera, I. N., Lipp, E. K., Gil, A., Choopun, N., Huq, A. & Colwell, R. R. (2003). Method of DNA extraction and application of multiplex polymerase chain reaction to detect toxigenic *Vibrio cholerae* O1 and O139 from aquatic ecosystems. *Environ. Microbiol.*, *5*, 599-606.

Rochelle, P. A., De Leon, R., Stewart, M. H. & Wolfe, R. L. (1997). Comparison of primers and optimization of PCR conditions for detection of *Cryptosporidium parvum* and *Giardia lamblia* in water. *Appl. Environ. Microbiol.*, *63*, 106-114.

Rousselon, N, Delgene,s, J. P. & Godo,n, J. P. (2004). A new real time PCR (TaqManR PCR) system for detection of the16S rDNA gene associated with fecal bacteria. *J. Microbiol. Methods*, *59*, 15-22.

Rudi, K., Moen, B., Drømtorp, S. M. & Holck, A. L. (2005a). Use of ethidium monoazide and PCR in combination for quantification of viable and dead cells in complex samples. *Appl. Environ. Microbiol.*, *71*, 1018-1024.

Rudi, K., Naterstad, K., Drømtorp, S. M. & Holo, H. (2005b). Detection of viable and dead *Listeria monocytogenes* on gouda-like cheeses by real-time PCR. *Lett. Appl. Microbiol.*, *40*, 301-306.

Saiki, R. K., Gelfand, D. H., Stoffel, S., Scharf, S. J., Higuchi, R., Horn, G. T., Mullis, K. B. & Erlich, H. A. (1988). Primer-directed enzymatic amplification of DNA with a thermostable DNA polymerase. *Science*, *239*, 487-91.

Samadpour, M. (2002). Microbial source tracking: Principles and practice, 5-10, *In: Microbiological Source Tracking Workshop-Abstracts*. February 5, 2002. Irvine, CA. NWRI Abstract Report NWRI-02-01. National Water Research Institute, Fountain Valley, CA.

Sambrook, J., Fritsch, E. F. & Maniatis, T. (2001). *Molecular cloning: a laboratorymanual*, 3rd ed. Cold Spring Harbor Laboratory Press, Cold Spring Harbor, N.Y.

Sandberg, M., Nygard, K., Meldal, H., Valle, P. S., Kruse, H. & Skjerve, E. (2006) Incidence trend and risk factors for Campylobacter infections in humans in Norway. *BMC Public Health*, *6*, 179-187.

Saundersa, A. M., Kristiansena, A., Lunda, M. B., Revsbecha, N. P. & Schramm, A. (2009). Detection and persistence of faecal Bacteroidales as water quality indicators in unchlorinated drinking water. *System. and Appl. Microbiol.*, In press.

Savichtcheva, O., Okayama, N., Ito, T. & Okabe, S. (2005). Application of a direct fluorescent-based live/dead staining combined with fluorescent in situ hybridization for assessment of survival rate of *Bacteroides* spp. in oxygenated water. *Biotechnol. Bioeng.*, *92*, 356-63.

Schets, F. M., van Wijnen, J. H., Schijven, J. F., Schoon, H. & de Roda Husman, A. M. (2008). Monitoring of waterborne pathogens in surface waters in Amsterdam, The Netherlands, and the potential health risk associated with exposure to *Cryptosporidium* and *Giardia* in these waters. *Appl. Environ. Microbiol.*, *74*, 2069-2078

Schonenbrucher, H., Abdulmawjood, A., Failing, K. & Bulte, M. (2008). A new triplex real time-PCR-assay for detection of *Mycobacterium avium* ssp. paratuberculosis in bovine faeces. *Appl. Environ. Microbiol.*, *74*, 2751-2758.

Schwab, K. J., De Leon, R. & Sobsey, M. D. (1995). Concentration and purification of beef extract mock eluates from water samples for the detection of enteroviruses, hepatitis A virus and Norwalk virus by reverse transcription-PCR. *Appl. Environ. Microbiol.*, *61*, 531-537.

Scott, T., Rose, J. B., Jenkins, T. M., Farrah, S. R. & Lukasik, J. (2002). Microbial source tracking: current methodology and future directions. *Appl. Environ. Microbiol.*, *68*, 5796-5803.

Selvaratnam, S., Schoedel, B. A., McFarland, B. L. & Kulpa, C. F. (1995). Application of reverse transcriptase PCR for monitoring expression of the catabolic *dmpN* gene in a phenol-degrading sequencing batch reactor. *Appl. Environ. Microbiol.*, *61*, 3981-3985.

Seurinck, S., Verstraete, W. & Siciliano, S. D. (2003). Use of 16S-23S rRNA intergenic spacer region PCR and repetitive extragenic palindromic PCR analyses of *Escherichia coli* isolates to identify nonpoint fecal sources. *Appl. Environ. Microbiol.*, *69*, 4942-4950.

Shangkuan, Y. H., Show, Y. S. & Wang, T. M. (1995). Multiplex polymerase chain reaction to detect toxigenic *Vibrio cholerae* and to biotype *Vibrio cholerae* O1. *J. Appl. Bacteriol.*, *79*, 264-273.

Sharma, V. K. & Dean-Nystrom, E. A. (2003). Detection of enterohemorrhagic *Escherichia coli* O157:H7 by using a multiplex real-time PCR assay for genes encoding intimin and Shiga toxins. *Veterinary Microbiol.*, *93*, 247-260.

Sheridan G. E. C., Masters, C. I., Shallcross, J. A. & Mackey B. M. (1998). Detection of mRNA by Reverse Transcription-PCR as an Indicator of Viability in *Escherichia coli* Cells. *Appl. Environ. Microbiol.*, *64*, 1313-1318.

Shirai, H., Nishibuchi, M., Ramamurthy, T., Bhattacharya, S. K., Pal, S. C. & Takeda, Y. (1991). Polymerase chain reaction for detection of the cholera enterotoxin operon of Vibrio cholerae. *J. Clin. Microbiol.*, *29*, 2517-2521.

Simpson, J. M., Santo Domingo, J. W. & Reasoner, D. J. (2002). Microbial source tracking: state of the science. *Environ. Sci. Technol.*, *24*, 5279-5288.

Singh, D. V., Matte, M. H., Matte, G. R., Jiang, S., Sabeena, F., Shukla, B. N., Sanyal, S. C., Huq, A. & Colwell, R. R. (2001). Molecular analysis of *Vibrio cholerae* O1, O139, non-O1, and non-O139 strains: clonal relationships between clinical and environmental isolates. *Appl. Environ. Microbiol.*, *67*, 910-921.

Skirrow, M. B. (1990). Foodborne illness: Campylobacter. *Lancet*, *336*, 921-923.

Sobek, J. M., Charba, J. F. & Foust, W. N. (1966). Endogenous metabolism of *Azotobacter agilis*. *J. Bacteriol.*, *92*, 687-690.

Speers, D. J. (2006) Clinical applications of molecular biology for infectious diseases. *Clin. Biochem. Rev.*, *27*, 39-51.

Steinhauserova, I., Ceskova, J., Fojtikova, K. & Obrovska, I. (2001). Identification of thermophilic Campylobacter spp. by phenotypic and molecular methods. *J. Appl. Microbiol.*, *90*, 470-475.

Stender, H., Broomer, A. J., Perry-O'Keefe, K. O. H., Hyldig-Nielsen, J. J., Sage, A. & Coull, J. (2001). Rapid detection, identification, and enumeration of *E. coli* cells in municipal water by chemiluminescent in situ hybridization. *Appl. Environ. Microbiol.*, *67*, 142-147.

Stender, H., Fiandaca, M., Hyldig-Nielsen, J. J. & Coull, J. (2002). Review article, PNA for rapid microbiology. *J. Microbiol. Methods*, *48*, 1-17.

Stender, H., Lund, K., Petersen, K. H., Rasmussen, O. F., Hoongmanee, P., Mifrner, H. & Godtfredsen, S. E. (1999). Fluorescence in situ hybridization assay using peptide nucleic acid probes for differentiation between tuberculous and nontuberculous mycobacterium species in smears of mycobacterium cultures. *J. Clin. Microbiol.*, *37*, 2760-2765.

Straub T. M., Pepper I. L. & Gerba C. P. (1995). Comparison of PCR and cell culture for detection of enteroviruses in sludge-amended field soils and determination of their transport. *Appl. Environ. Microbiol.*, *61*, 2066-2068.

Studer, E. & Candrian, U. (2000). Development and validation of a detection system for wild-type Vibrio cholerae in genetically modified cholera vaccine. *Biologicals*, *28*, 149-154.

Suau, A., Bonnet, R., Sutren, M., Godon, J. J., Gibson, G. R., Collins, M. D. & Dore, J. (1999). Direct analysis of genes encoding 16S rRNA from complex communities reveals many novel molecular species within the human gut. *Appl. Environ. Microbiol.*, *65*, 4799-4807.

Suzuki, M., Rappe, M. S. & Giovannoni, S. J. (1998). Kinetic bias in estimates of coastal picoplankton community structure obtained by measurements of small-subunit rRNA gene PCR amplicon length heterogeneity. *Appl. Environ. Microbiol.*, *64*, 4522-4529.

Szewzyk, U., Szewzyk, R., Manz, W., Schleifer, K. H. (2000) Microbiological safety of drinking water. *Annu. Rev. Microbiol.*, *54*, 81-127.

The world health report 2001 - Mental Health: New Understanding, New Hope

Tomaso, H., Scholz, H. C., Neubauer, H., Al Dahouk, S., Seibold, E., Landt, O., Forsman, M. & Splettstoesser, W. D. (2007). Real-time PCR using hybridization probes for the rapid and specific identification of *Francisella tularensis* subspecies tularensis. *Mol. Cell Probes*, *21*, 12-16.

Tsai, Y. L. & Olson, B. H. (1990). Effects of Hg21, CH3-Hg1, and temperature on the expression of mercury resistance genes in environmental bacteria. *Appl. Environ. Microbiol.*, *56*, 3266-3272.

Tsen, H. Y., Lin, C. K. & Chi, W. R. (1998). Development and use of 16S rRNA gene targeted PCR primers for the identification of *Escherichia coli* cells in water. *J. Appl. Microbiol.*, *85*, 554-560.

van der Giessen, J. W., Eger, A., Haagsma, J., Haring, R. M., Gaastra, W. & van der Zeijst, B. A. (1992). Amplification of 16S rRNA sequences to detect *Mycobacterium paratuberculosis*. *J. Med. Microbiol.*, *36*, 255-263.

Van der Meide, P. H., Vijgenboom, E., Talens, A. & Bosch, L. (1983). The role of EF-Tu in the expression of *tufA* and *tufB* genes. *Eur. J. Biochem.*, *130*, 397-407.

Van der Vliet, G. M. E., Schepers, P., Schukkink, R. A. F., Van Gemen, B. & Klatser, P. R. (1994). Assessment of mycobacterial viability by RNA amplification. *Antimicrob. Agents Chemother.*, *38*, 1959-1965.

Vandenberg, O., Dediste, A., Houf, K., Ibekwem, S., Souayah, H., Cadranel, S., Douat, N., Zissis, G., Butzler, J. P. & Vandamme, P. (2004). *Arcobacter* species in humans. *Emerg. Infect. Dis.*, *10*, 1863-1867.

Versalovic, J., Schneider, M., de Bruijn, F. J. & Lupski, J. R. (1994). Genomic fingerprinting of bacteria using repetitive sequence-based polymerase chain reaction. *Methods Mol. Cell. Biol.* *5*, 25-40.

Vesper, S., McKinstry, C., Hartmann, C., Neace, M., Yoder, S. & Vesper, A. (2008). Quantifying fungal viability in air and water samples using quantitative PCR after treatment with propidium monoazide (PMA). *J. Microbiol. Methods*, *72*, 180-184.

Volokhov, D., Rasooly, A., Chumakov, K. & Chizhikov, V. (2002). Identification of *Listeria* species by microarray-based assay. *J. Clin. Microbiol.*, *40*, 4720-4728.

Vora, G. J., Meador, C. E., Bird, M. M., Bopp, C. A., Andreadis, J. D. & Stenge, D. A. (2005). Microarray-based detection of genetic heterogeneity, antimicrobial resistance, and the viable but nonculturable state in human pathogenic *Vibrio* spp. *PNAS*, *52*, 19109-19114.

Vora, G. J., Meador, C. E., Stenger, D. A. & Andreadis, J. D. (2004). Nucleic acid amplification strategies for DNA microarraybased pathogen detection. *Appl. Environ. Microbiol.*, *70*, 3047-3054.

Waage, A. S., Vardund, T., Lund, V. & Kapperud, G. (1999a). Detection of low numbers of *Salmonella* in environmental water, sewage and food samples by a nested polymerase chain reaction assay. *J. Appl. Microbiol.*, *87*, 418-428.

Waage, A. S., Vardund, T., Lund, V. & Kapperud, G. (1999b). Detection of low numbers of pathogenic Yersinia enterocolitica in environmental water and sewage samples by nested polymerase chain reaction. *J. Appl. Microbiol.*, *87*, 814-821.

Wang, X., Jothikumar, N. & Griffiths, M. W. (2004). Enrichment and DNA extraction protocols for the simultaneous detection of *Salmonella* and *Listeria monocytogenes* in raw sausage meat with multiplex real-time PCR. *J. Food Prot.*, *67*, 189-192.

Wang, Y., Hammes, F., Boon, N. & Egli, T. (2007). Quantification of the filterability of freshwater bacteria through 0.45, 0.22 and 0.1 mm pore size filters and shape-dependent enrichment of filterable bacterial communities. *Environ. Sci. Technol.*, *41*, 7080-7086.

Wassenaar, T. M. & Newell, D. G. (2000). Genotyping of *Campylobacter* spp. *Appl. Environ. Microbiol.*, *66*, 1-9.

Watson, C. L., Owen, R. J., Said, B., Lai, S., Lee, J. V., Surman-Lee, S. & Nichols, G. (2004). Detection of *Helicobacter pylori* by PCR but not culture in water and biofilm samples from drinking water distribution systems in England. *J. Appl. Microbiol.*, *97*, 690-698.

Way, J. S., Josephson, K. L., Pillai, S. D., Abaszadegan, M., Gerba, C. P. & Pepper, I. L. (1993). Specific detection of Salmonella spp. By multiplex polymerase chain reaction. *Appl. Environ. Microbiol.*, *59*, 1473-9.

Weinbauer, M. G., Fritz, I., Wenderoth, D. F., Hofle, M. G. (2002) Simultaneous extraction of total RNA and DNA from bacterioplankton suitable for quantitative structure and function analyses. *Appl. Environ. Microbiol.*, *68*, 1082-1087.

Wellinghausen, N., Frost, C. & Marre, R. (2001). Detection of *Legionellae* in hospital water samples by quantitative real-time LightCycler PCR. *Appl. Environ. Microbiol.*, *67*, 3985-3993.

White, T. J. (1996). The future of PCR technology: diversification of technologies and applications. *Trends Biotechnol.*, *14*, 478-483

Wilkes, G., Edge, T., Gannon, V., Jokinen, C., Lyautey, E., Medeiros, D., Neumann, N., Ruecker, N., Topp, E. & Lapen, D. R. (2009). Seasonal relationships among indicator bacteria, pathogenic bacteria, *Cryptosporidium oocysts*, *Giardia cysts*, and hydrological indices for surface waters within an agricultural landscape. Water research, In press.

Wilson, K. H. & Blitchington, R. B. (1996). Human colonic biota studied by ribosomal DNA sequence analysis. *Appl. Environ. Microbiol.*, *62*, 2273-2278.

Wilson, W. J., Strout, C. L., DeSantis, T. Z., Stilwell, J. L., Carrano, A. V. & Andersen, G. L. (2002). Sequence-specific identification of 18 pathogenic microorganisms using microarray technology. *Mol. Cell. Probes*, *16*, 119-127.

World Health Organization (WHO). 1996. World health report 1996: fighting disease, fostering development. Geneva: World Health Organization.

World Health Organization (WHO). 2001. World health report 2001: Mental Health: New Understanding, New Hope. Geneva: World Health Organization.

World Health Organization (WHO). 2003. World health report .2003. Emerging issues in water and infectious disease. Geneva: World Health Organization.

World Health Organization (WHO). 2004. World health report 2004: Changing history. Geneva: World Health Organization.

Xiao, L., Singh, A., Limor, J., Graczyk, T. K., Gradus, S. & Lal, A. (2001). Molecular characterization of *Cryptosporidium oocysts* in samples of raw water and wastewater. *Appl. Environ. Microbiol.*, *67*, 1097-1101.

Yang, Y. G., Song, M. K., Park, S. J. & Kim, S. W. (2007). Direct detection of *Shigella flexneri* and *Salmonella typhimurium* in human feces by real-time PCR. *J. Microbiol. Biotechnol.*, *17*, 1616-1621.

# INDEX